Food and Identity in a Globalising World

Series Editors
Atsuko Ichijo
Department of Politics
Kingston University
Kingston-Upon-Thames, UK

Ronald Ranta
Department of Politics
Kingston University
Kingston-Upon-Thames, UK

This series aims to overcome the current fragmented nature of the study of food by encouraging interdisciplinary studies of food and serving as a meeting place for a diverse range of scholars and practitioners who are interested in various aspects of food. By encouraging new original, innovative and critical thinking in the field and engaging with the main debates and controversies, and by bringing together the various disciplines that constitute food studies, such as, sociology, anthropology, politics and geography, the series will serve as a valuable source for researchers, practitioners, and students. There will a focus on identities and food; issues such as gastrodiplomacy, settler colonialism, gender, migration and diaspora, and food and social media, while at the same time promoting an inter- and trans-disciplinary approach.

More information about this series at
https://link.springer.com/bookseries/16371

Ronald Ranta • Alejandro Colás
Daniel Monterescu
Editors

'Going Native?'

Settler Colonialism and Food

palgrave
macmillan

Editors
Ronald Ranta
Department of Politics and International Relations
Kingston University
Kingston upon Thames, UK

Alejandro Colás
Department of Politics
Birkbeck, University of London
London, UK

Daniel Monterescu
Department of Sociology and Social Anthropology
Central European University
Vienna, Austria

ISSN 2662-270X ISSN 2662-2718 (electronic)
Food and Identity in a Globalising World
ISBN 978-3-030-96267-8 ISBN 978-3-030-96268-5 (eBook)
https://doi.org/10.1007/978-3-030-96268-5

Cover credit: Witold Skrypczak / Alamy Stock Photo

This Palgrave Macmillan imprint is published by the registered company Springer Nature Switzerland AG.
The registered company address is: Gewerbestrasse 11, 6330 Cham, Switzerland

Contents

1 **Introduction** 1
Alejandro Colás, Daniel Monterescu, and Ronald Ranta

Part I Beginnings: Hybrid Food Cultures and Foodways 21

2 **Spanish Settlers and Andean Food Systems** 23
Lisa Markowitz

3 **What Belongs in the "Federal Diet"? Depictions of a National Cuisine in the Early American Republic** 47
Peter Mabli

4 **The Taste of Colonialism? Changing Norms of Rice Production and Consumption in Modern Taiwan** 65
Yujen Chen

5 **'Like the Papacy of Mexican Cuisine':** ***Mayoras*** **and Traditional Foods in Contemporary Mexico** 85
Claudia Prieto-Piastro and Alejandro Colás

Part II From Erasure to Decolonisation 107

6 **Unsettling the History of Macadamia Nuts in Northern New South Wales** 109
Adele Wessell

7 **Definitions of Hawaiian Food: Evidence of Settler Colonialism in Selected Cookbooks from the Hawaiian Islands (1896–2021)** 127
Laura Kitchings

8 **Decolonising Israeli Food? Between Culinary Appropriation and Recognition in Israel/Palestine** 147
Ronald Ranta and Daniel Monterescu

9 **'Uneasy Lies the Head That Wears the Crown': Lamb or Kangaroo, Which Should Reign Supreme? The Implications of Heroising a Settler Colonial Food Icon as National Identity** 173
Jacqueline Newling

Part III After Decolonisation? 201

10 **"A Manly Amount of Wreckage": South African Food Culture and Settler Belonging in Ivan Vladislavić's *Double Negative*** 203
Nitzan Tal

11 **Sustaining the Memory of Colonial Algeria Through Food** 221
Amy L. Hubbell and Jorien van Beukering

12 **The Predicaments of Settler Gastrocolonialism** 247
Lorenzo Veracini

Index 261

Notes on Contributors

Yujen Chen is Professor of Taiwan Culture, Languages, and Literature at the National Taiwan Normal University in Taiwan. From 2016 to 2018, she was an adjunct research fellow of KEIO Institute of East Asian Studies in Japan. Chen holds a PhD degree (Faculty of Humanities) from Leiden University in the Netherlands. Chen's main research interests include food history in Taiwan and East Asia, media and consumption society, food anthropology, culinary literature, and interdisciplinary research methodology. She is the author of *The Cultural History of 'Taiwanese Cuisine'* (2020). Chen also published several research articles in journals such as *Food, Culture and Society*, *Taiwan Journal of Anthropology*, *Journal of Taiwan History*, and *Journal of Family History*. Her next book project will be the history of the dairy industry in Taiwan.

Alejandro Colás is Professor of International Relations in the Politics Department at Birkbeck College, University of London. He is co-author of *Food, Politics and Society: Social Theory and the Modern Food System* (2018).

Amy L. Hubbell is Senior Lecturer in French at The University of Queensland where she researches Francophone autobiographies of exile and trauma. She is an author of *Hoarding Memory: Covering the Wounds of the Algerian War* (2020) and *Remembering French Algeria: Pieds-Noirs, Identity and Exile* (2015) and has co-edited several volumes of essays including *Places of Traumatic Memory—A Global Context* (2020), *The Unspeakable: Representations of Trauma in Francophone Literature and Art* (2013), and *Textual and Visual Selves: Photography, Film and Comic Art in French Autobiography* (2011).

Laura Kitchings is a cultural heritage professional and holds Master of Arts in Museology from University of Washington and Master of Science in Library Science and Master of Arts in History from Simmons College (now Simmons University). She has worked as an archivist for a number of organisations in Massachusetts. In 2021, she was the recipient of the Metropolitan College Award for Excellence in Graduate Studies in Gastronomy from Boston University when she was awarded her Master of Liberal Arts in Gastronomy. Her research focuses on interdisciplinary methodologies for analysing the use of food in archival sources.

Peter Mabli serves as Director of Online Programs at the American Social History Project, a public history organisation based at the City University of New York. He is also Professor of History at Fairleigh Dickinson University, where he develops and teaches online courses in US history and the history of New Jersey. Peter holds a PhD in American History and Culture (2019) from Drew University, with a focus on the intersections between food culture and national identity formation in the early American Republic.

Lisa Markowitz (University of Louisville) is a cultural anthropologist interested in agrifood systems at multiple scales and popular efforts to transform them. Since the 1980s she has carried out fieldwork in the Andean region with foci on rural livelihood, food security, and value chains. She has also been engaged in research with and about food activists in the United States. Markowitz is co-editor of *U.S. Food Policy: Anthropology and Advocacy in the Public Interest* (2012) and a former President of the Culture & Agriculture Section of the American Anthropological Association.

Daniel Monterescu is Associate Professor of Urban Anthropology and Food Studies at the Central European University, Vienna. He holds a PhD from the University of Chicago, a Sommelier certificate (level 3, 2010, Italy) and is training for completion of the Wine & Spirit Education

Trust level 4 diploma in Austria. He studies food politics and border wines in Europe and the Middle East through the concepts of terroir and territory. Most recently, he is co-author of 'Soil, Territory, Land: The Spatial Politics of Settler Organic Farming in the West Bank' (*Environment and Planning D: Society and Space*, 2021).

Jacqueline Newling is a historian, gastronomer, museum curator, Honorary Associate in History at the University of Sydney, and an accredited Professional Historian. She lives and works on Wangal and Gadigal Country, unceded lands of Australia's First Nations people. She is a graduate of the Le Cordon Bleu Masters in Gastronomy program, and has a specialisation in Australian settler-colonial food culture. She is the author of *Eat Your History, Stories And Recipes From Australian Kitchens, 1788–1950s* (2015).

Claudia Prieto-Piastro is an Anthropologist and the Programme Element Leader for Social Sciences, Humanities, and Education at Brunel University London Pathway College. Her research focuses on everyday nationalism, creolisation, and the role of food in the construction of national identity as well as in immigration, food culture, and religious identities.

Ronald Ranta is a former chef and Senior Lecturer in Politics and International Relations at Kingston University. His research focuses on the intersection of food, identity, and politics. He is the author of *Food, National Identity and Nationalism* (2016) and *From the Arab Other to the Israeli Self: Palestinian Culture in the Masking of Israeli National Identity* (2016).

Nitzan Tal received her PhD from the Department of Comparative Literature at Cornell University. Her work interrogates contemporary Hebrew and South-African novels as sites for the adjudication of settler-colonial complicity and of theories of social responsibility. Her work has been published in *Middle Eastern Literatures* and in the *Journal of Modern Jewish Studies*. Her Hebrew translation of Sara Ahmed's "Feminist Killjoys" is available as part of her co-edited feminist theory translation series.

Jorien van Beukering is a PhD student and research assistant in the School of Languages and Cultures at The University of Queensland. Her research analyses identity in illegitimate children of biracial heritage with a familial link to the Dutch East Indies (colonial Indonesia). Her broader research interests include French and Francophone culture and history, Dutch (post)colonial history, memory, trauma, and identity studies.

Lorenzo Veracini teaches history and politics at Swinburne University of Technology, Melbourne. His research focuses on the comparative history of colonial systems and settler colonialism as a mode of domination. He has authored *Israel and Settler Society* (2006), *Settler Colonialism: A Theoretical Overview* (2010), *The Settler Colonial Present* (2015), and *The World Turned Inside Out* (2021). Veracini co-edited *The Routledge Handbook of the History of Settler Colonialism* (2016), manages the settler colonial studies *blog*, and was Founding Editor of *Settler Colonial Studies.*

Adele Wessell is Associate Professor of History at Southern Cross University, based in the Northern Rivers region, the traditional lands of the Bundjalung people. Wessell is a food historian and co-editor of *Locale: Pacific Journal of Regional Food Studies*.

List of Figures

Fig. 3.1 Moyne, Jacques le. *Plate XXIX: A Council of State*. 1564. Engraving. Special Collections Department. University of South Florida Archives 54

Fig. 9.1 'Sheep & Wheat' c.1920s map of Australia showing the extent of land used for sheep and wheat growing (Philip et al. c.1920) (George Philip and Son, and Taylor, Thomas Griffith, and Beckit, H.O. National Library Australia http://nla.gov.au/nla.obj-234331946. An equivalent 'Cattle and Minerals' map (National Library of Australia MAP G8961.J1 [192–?] https://nla.gov.au/nla.obj-255201355/view) adds to the picture. For more current distribution of sheep stocks, see MLA map 'Sheep numbers as at June 30 2019' (Accessed March 30, 2021). https://www.mla.com.au/globalassets/mla-corporate/prices%2D%2Dmarkets/documents/trends%2D%2Danalysis/fast-facts%2D%2Dmaps/2020/mla-sheep-numbers-map-2020-at-june-2019.pdf) 186

Fig. 9.2 'Tindale's Arc' (Chivers et al. 2015, 6, following Tindale, 1974) showing grassland areas traditionally used by Aboriginal people for grain. According to Chivers et al., other evidence suggests that grain processing was 'much more widespread than suggested in [Tindale's] map' (2015, 5, 6). Viewed in parallel with Fig. 9.1, we see the potential impact of sheep on

Aboriginal people's traditional foodscape, in Australia's eastern and western regions. 187
Fig. 9.3 Livestock grazing pressures 1888–2008 (ABS) 2010, n.p.). According to the ABS, 'agriculture is the most extensive form of land use in Australia [and] livestock grazing accounts for the largest area of land use in agriculture.' (2010, n.p) 188

1

Introduction

Alejandro Colás, Daniel Monterescu, and Ronald Ranta

> *'As settlers are non-indigenous, they are forever indigenising ... The question then is how to be efficient self-indigenisers'*
> —Veracini (*2015*: 270)

What do settler colonial foodways and food cultures look like? Are they based on an imagined colonial heritage, do they embrace indigenous repertoires or invent new hybridised foodscapes? What are the

A. Colás
Department of Politics, Birkbeck, University of London, London, UK
e-mail: a.colas@bbk.ac.uk

D. Monterescu
Department of Sociology and Social Anthropology, Central European University, Vienna, Austria
e-mail: monterescud@ceu.edu

R. Ranta (✉)
Department of Politics and International Relations, Kingston University, Kingston upon Thames, UK
e-mail: ronald.ranta@kingston.ac.uk

R. Ranta et al. (eds.), *'Going Native?'*, Food and Identity in a Globalising World,
https://doi.org/10.1007/978-3-030-96268-5_1

socio-economic and political dynamics of these cultural transformations? This volume offers a comparative survey of diverse settler colonial experiences in relation to food, food culture and foodways—how the latter are constructed, maintained, revolutionised and, in some cases, dissolved. In particular, this volume focuses on three key issues: the evolution of settler colonial identities and states; their relations vis-à-vis indigenous populations; and settlers' self-indigenisation—the process through which settlers transform themselves into the native population, at least in their own eyes. These three key issues are crucial in understanding the rise of settler colonial identities and states and their interaction with the indigenous populations that inhabit them.

In recent years two parallel culinary debates have occurred in many settler colonial states regarding indigenous people.[1] The first debate concerns the growing interest in indigenous food cultures, foodways, terroir and culinary heritage (e.g. see Banerjee-Dube 2016; Grey and Newman 2018; and Monterescu and Handel 2019). Indigenous foods, which have been, in some cases, previously neglected or rejected, have been 'rediscovered', celebrated and occasionally incorporated or appropriated as part of the national settler colonial food culture. This 'rediscovery' is sometimes presented as part of a decolonising move towards liberal cosmopolitan societies, which are more inclusive of indigenous people and their cultures (Grey and Newman 2018). The growing interest is also part of a wider indigenous food movement that emphasises indigenous food sovereignty (e.g. see Daigle 2019). This is happening in parallel to a debate concerning the place and rights of indigenous people and at times other minorities, within settler colonial states. In terms of food, more inclusive and/or multicultural sensibilities promote indigenous rights to food sovereignty and to practice traditional ways of life, including hunting, fishing and foraging, and access to their land, water and related resources.

The above debates raise multifaceted and difficult questions over the main drivers and beneficiaries of this renewed interest in indigenous food

[1] The term indigenous has been and is mostly used to refer to the descendants of the pre-colonial peoples of several regions, including the Americas, the Arctic territories, Australia and New Zealand. The term is often used interchangeably with the terms first people, aboriginal people, native people or autochthonous people. In addition, other, often marginalised, ethnic and national groups with distinct cultures, define themselves as indigenous.

and whether these debates signify a shift in thinking regarding the nature and structure of settler colonial identities and states. One major concern is that such fresh attention on indigenous foodways ends up commodifying and ossifying these cultural forms into reified spectacles and performances, rather than opening up new spaces of recognition and creativity. This concern is in line with Veracini's argument that 'privileging the indigenous over the non-indigenous' may turn colonial logic upside down without necessarily changing it (2011: 7–8). As Adele Wessel points out in her chapter on macadamia nuts, 'embracing native foods can be used in the construction of a national cuisine, an imagined connection and belonging to a place still predicated in erasing Indigenous food sovereignty and appropriation' (this volume, 123).

Before delving further into the above issues and debates, it is important to clarify what this volume does and does not do. The volume is a critical, multidisciplinary and comparative study of settler colonial identities and states: it is not an attempt to provide an indigenous perspective on these issues or an account of indigenous food cultures and foodways. All the scholars in this volume are either descendants of settlers or part of the critical settler colonial paradigm, researching and writing as 'insiders' about settler colonial identities and states. We would like to make clear that we are not speaking on behalf of indigenous people or trying to represent their views; in cases where indigenous views are provided these are based on either primary sources or indigenous scholarship; though we fully acknowledge that much of the primary evidence regarding the food culture, foodways, food perspectives and experiences of indigenous people was generally documented and interpreted by settlers.

The focus of this volume is on settler colonial foodways and food cultures rather than indigenous ones, though the balance of analysis is different in each chapter. Nevertheless, it is clear that indigenous foodways and food cultures have been profoundly altered by settler colonialism, through, for example, the forceful transformation of the land, the removal of indigenous people from it, and the introduction of new crops and animals. Historically, colonised and occupied people and cultures have struggled to resist the encroachment of cultural materialism and the structures, changes and ideas, which were in many cases imposed upon them (Ranta and Mendel 2016; Wilk 2006). There has been substantial research on the impact of settler colonialism on the cultures and identities of indigenous people, much of it from an indigenous perspective (e.g. see Bhabha 1997; Fanon 1991).

What Is Settler Colonialism? A Relational Approach

The term settler colonialism is used here to refer to the movement of people from one place, normally from the metropole[2] and often with the support of an imperial power, for the purpose of permanently settling in and occupying another. It differs from associated terms, such as colonialism and imperialism, in that settler colonialism is not specifically about the exploitation of the indigenous people it encounters or the land and its resources, though these are more often than not associated with the process, but about the acquisition of and taking over the land for the purpose of settling it. This process of acquiring and settling the land normally leads, by design or by default, either to the marginalisation of the indigenous people who previously inhabited the land or to their forceful removal from it. Thus the relationship between the settler and the indigenous is always in relation to issues of land and power. In contrast to forms of colonialism that seek to perpetuate their power mechanisms, settler colonialism aims to supersede, normalise and exhaust the colonial situation: 'a settler colonial project that has successfully run its course is no longer settler colonial' (Veracini 2013: 29).

In trying to understand the evolution and construction of settler colonial identities and states the volume focuses on the process of settler self-indigenisation and on relations vis-à-vis the indigenous populations. Reframing a structure of power as a historical process, we propose to read the relations between settlers and indigenous identities as a relational unfolding process—problematising Wolfe's argument (2006) that the settlers came to stay, thus viewing settler colonialism as a structure rather

[2] Though the term is normally used to denote the metropolitan centre of diverse Empires, we used it here more broadly to denote the political and cultural centres of imperial power (the obvious exception here is Israel—a settler colonial state without a clear cut metropole).

than an event. Self-indigenisation is the process through which settlers seek to turn themselves into the native population and to naturalise their very existence in the land of contention. Food, including drink, in this context becomes an emblematic vehicle for settlers to write themselves into and shape the territory and history; the process also directly questions who and what should be considered native and/or indigenous and part of the settler constructed state. As Mamdani (1998) notes: 'the settler-native question is a political question. It is also a historical question. Settlers and natives belong together. You cannot have one without the other, for it is the relationship between them that makes one a settler and the other a native. To do away with one, you have to do away with the other'. In short, the ways in which the category of the indigenous is discursively framed varies between cases.

Settler colonialism is primarily about power and land. It is the domination of one group, albeit a heterogeneous one, and its use of appropriated indigenous lands primarily for the production of food to sustain its polity. Food is thus central to the establishment and survival of the settler colony and settler colonialism. Food, however, is not merely about consumption and sustenance. Food is a marker of identity in that what people eat informs 'who and what they are, to themselves and to others' (Mintz 1985: 13); this point is relevant not only at an individual level, but also on a societal and national one (Ranta and Ichijo 2016). Food is also embodied material culture and thus reflects the politics of its production, reception and consumption. Linking the above points, food can provide a useful prism through which to examine and think of the formation, evolution and maintenance of group and national identities. As noted by Bell and Valentine: 'the history of any nation's diet is the history of the nation itself, with food fashions, fads and fancies mapping episodes of colonialism and migration, trade, and exploration, cultural exchange and boundary-marking' (1997: 168–169). In short, the study of settler foodways and food cultures provide us with a window, for understanding not only how settler colonies are formed and evolve, but also the evolution of relations and power dynamics within settler societies and between the settlers and the indigenous populations. As Lisa Markowitz explains in her chapter on Peru, the 'examination of contemporary [settler] foodways reveals deeper histories and connections' (this volume, 25).

The use of food by the settlers is multifaceted, but we would like to highlight three areas in particular. First, settlers adopt, adapt and appropriate indigenous food traditions and knowledge when transforming themselves into the 'new' natives. As the dominant power, settlers have been mostly able to choose the elements they want to adopt and those they prefer to reject from the indigenous foodways and food cultures (Wilk 2006). Second, the process of self-indigenisation is not based solely on the adoption, adaptation and appropriation of indigenous traditions and knowledge, but also based on the transformation of the land and the introduction of new crops, animals and farming methods brought from their former homelands. Third, the creation of the new settler societies, based on the hybridisation of indigenous and former traditions and knowledge, is supported by the displacement of the indigenous populations and their crops, farming and hunting, often generating new settler colonial expressions of landscape and terroir. As Claudia Prieto Piastro and Alejandro Colás explain in their chapter on Mexico, settler colonial foodways and food culture need to be understood and situated within a 'wider socio-economic and political context' that acknowledges 'the structural legacies … relating to land tenure, collective rights or racial hierarchies' (this volume, 104).

In focusing on settler self-indigenisation and the relationship between settlers and indigenous populations, the volume asks four pertinent questions. First, and thinking of these issues historically, whether and to what extent is the process of self-indigenisation based on the incorporation, adaptation and appropriation of indigenous culture and knowledge, in our case food culture? Second, how do settler self-indigenisation and settler-indigenous relations correspond to different modalities of settler colonial relations—domination, control, erasure, co-optation, appropriation, integration, mimicry, hybridity and resistance? Third, to what extent is the renewed interest in indigenous foodways and food cultures related to processes of decolonisation? Lastly, what happens to self-indigenising and the separate settler identity after successful decolonisation?

Self-indigenisation

When thinking about settler colonial identities and states, it is possible to broadly discuss a starting point, a moment in time in which the process began: the moment settlers arrived; though it is clear that settlers continue 'to arrive' from that point onwards. The settlers rather than immigrate to another country, immigrate, either by design or by default, to establish their 'own' country and create their 'own' nation (Veracini 2010). The arrival of settlers, and their desire to settle and transform, what they see as, 'their' land and create their 'own' nation, is, in most cases, an existential threat to indigenous populations. In terms of the relationship between the settlers and the indigenous populations, we do not dispute Wolfe's (2006) assertion that settler colonialism destroys the existing indigenous social and political structures in order to replace them with its own. However, as the volume demonstrates, there is more to the settler-indigenous relations. Wolfe (1999) argues that the relationship is not so much between settler colonialists and the indigenous societies they encounter as between the settlers and the land. The settlers' need for more land eventually necessitates the removal of the indigenous population and in some cases their outright extermination. As we demonstrate, settler-indigenous relations were and are not simply about the elimination of the indigenous people and their culture to replace them with a settler society, although we fully acknowledge that this is often a feature of settler colonialism. In addition to acquiring land, settler colonialists also undergo a process of change that has implications for their relations with the indigenous populations: transforming themselves into the native inhabitants by establishing political frames of representation and forging new discourses of rights. This transculturation or self-indigenisation manifests differently in each settler society, depending on their relations with the indigenous populations and the interactions between existing settlers and new arrivals, including indentured workers and slaves, with examples of hybrid and creole or mestizo identities. These complex cultures are particularly evident in long-established settler colonial states, such as in the Americas, as demonstrated in the chapters on Mexico and Peru. Here it is also worth noting the cleavages of class, nationality and indeed gender that cut across the otherwise homogenised category of the 'settler': in colonial North Africa and elsewhere a stark social distinction between the 'petits blans' (transport workers, dockers, office clerks) and the 'grand families' of settler capitalism, as did the differentiation between, say, French settlers and later Spanish, Italian or Maltese immigrant arrivals; this issue is further explored in the Algerian case study chapter.

It is therefore important not to reproduce the impression of a monolithic or uniform settler society or identity. As we explained above, settlers are a diverse category that often includes a range of identities, including nationalities, ethnicities and religions, and their accompanying complex power relations. Within the settler category are marginalised and oppressed groups, including indentured and enslaved workers, as well as racialised white settlers, such as the Irish and Jews in the case of the US. When thinking about settler foodways and food cultures, it is thus important to recognise the complexity of the settler category and the multi-layered social relations that exist between the various groups that constitute it, as well as between settlers and indigenous people (e.g. see Gabaccia 1998; and Harris 2012; on the entangled relationships between indigenous people, enslaved workers, migrants and white elites, in the construction of American foodways and food culture).

The complexity of settler identities and societies means that the process of self-indigenisation occurs at several different levels. Settlers transform themselves into the natives in relation to the indigenous people and cultures they encountered, which includes the diffusion and transfer, often by force, of ideas and knowledge. This fits in with Hixson's (2013) argument relating to the existence of a range of interactions between settlers and indigenous people in North America. Parts of the indigenous culture and knowledge—those that are of use to the settlers—are adopted, adapted and appropriated by the settlers, either for instrumental reasons or as part of the process of transforming themselves into the new natives. In Israel/Palestine, for instance, Zionist settlers adopted local plants, such as the prickly pear and the orange, as their national symbols to emphasise their nativeness (Abufarha 2008; and Bardenstein 1998). As discussed in his chapter on the Federal diet, Peter Mabli provides examples of how American settlers were selective in their adoption of indigenous food, for example, rejecting jappon tea as unfit for consumption and naming it, unfortunately, *Ilex vomitoria* (this volume, 56).

After the settler colonial state is established, settler foodways and food culture go through a process of de-indigenising. The term is used here to refer to the gradual process by which settler colonialists rewrite and forget their history, denying the vital influence of indigenous foodways and food cultures and marginalising it in the construction of the new settler identity. The process begins when the settlers' desire for more land and a separate settler state increases and the conflict between the settlers and indigenous population intensifies. Indigenous food cultures and foodways are then presented as part of the founding myths, histories, customs and practices of the settler nation, and after the creation of the settler nation-state, these are often nationalised. Settler national symbols also include food items that they have introduced, but devoid of their colonial history, as in the case of lamb in Australia and dairy milk in North America serve as national food symbols. This process allows for the construction of a national narrative that charts the transformation of the settlers into the natives, but which does not include the contribution made by indigenous people and culture. As many of the volume's case studies demonstrate, settler authenticity and rootedness are often predicated on the logic of elimination, silencing and institutional forgetting, with national history replacing settler violence. A point emphatically made by Aboriginal Australian food scholar Bruce Pascoe: 'You can't eat our food if you can't swallow our history' (Allemann 2018).

The settler colonial encounter was never unidirectional. It always involved cross-references, rejection and appropriation. Our proposed revisionist conceptualisation of the settler colonial encounter makes visible, as Albert Memmi (1985) has noted, the dialectic enchaînement between the coloniser and the colonised that produces in the process multiple intentionalities, identifications and alienations. The culinary colonial frontier thus emerges not as a site of zero-sum conflict but rather as 'a place in which the unfolding histories' of both dominators and dominated, centre and periphery, 'met—there to be made, reciprocally, in relation to each other'. Beyond and beneath colonialism's black-and-white dualisms and 'working essentialisms,' the settler-indigenous encounter did create an 'awareness of ruptures at which local resistance was directed, and in which new hybridities could take root' (Comaroff and Comaroff 2009, 403).

Methodology and Structure

As we explain above, this volume examines settler colonial foodways and food cultures, with a focus on their evolution, relationships vis-à-vis indigenous populations and their processes of self-indigenisation. It was clear to us, when we first conceived of this volume, that the most appropriate approach for analysing the subject would be a comparative and holistic one, which would enable us to capture the various and complex dimensions of the subject. In order to do so, we predicated our case study selection on examples that span different time periods and geographic areas and include a range of disciplinary and multidisciplinary approaches from the social sciences and humanities.

The cases selected provide examples from three crucial time periods, namely the establishment and evolution of settler colonial societies and states; established settler colonial states debating cosmopolitanism and decolonisation; and the ongoing and at times painful process of decolonisation which defines the postcolonial condition in the present and the projected future. In terms of geography, we included cases from the Americas, Africa and the Middle East, Asia, and Australia; we also included one case study focused on a specific food item, namely macadamia nuts. We left out only one region, partly by choice and partly by default. We had discussions with several potential contributors regarding a European case study—Ireland and Northern Ireland—but for various reasons these did not pan out. Nevertheless, the centrality of Europe and European societies and states to settler colonialism is discussed extensively in all of the chapters, including that of Taiwan.

In terms of methodology, food as a field of study naturally lends itself well to a holistic and multi- and cross-disciplinary approach (Mintz and Du Bois 2003). We were conscious that simply following a historical or sociological route would not be sufficient to capture the complexity and diversity of settler colonial foodways and food cultures. The approach taken, to include a range of methodological and disciplinary viewpoints, helps us bring into the volume historical as well as political-economic and socio-cultural elements. It allows the volume to include a wide range of important issues, such as race, ethnicity, ecology, nationalism, migration, rights and justice. Additionally, the diversity of case studies and methodologies helps us analyse the phenomenon from different angles of settler foodways and food cultures, from land clearing and cultivation to consumption and commodification.

Comparative edited volumes too often end up as collections of ad hoc chapters with little synergy between them. In order to mitigate this problem, we decided to invite the case study contributors to be part of the process of editing the volume. We circulated all of the chapters, including the introduction, amongst the contributors and held two writing and feedback workshops. This provided contributors the space to discuss their own work, but also offer feedback on other case studies and on the volume, and we made a concerted effort to include points made by the contributors. In doing so we tried to use the volume to create a platform for genuine dialogue between scholars studying this subject.

This volume was written in the midst of the COVID-19 pandemic, and we acknowledge how difficult it has been for contributors to participate in the workshops and submit their chapters on time given the challenges involved, including lockdowns, home-schooling and, in some cases, illness. Nevertheless, the volume is not as complete as we initially intended it to be. At different stages, several authors who agreed to contribute had to withdraw, with the result that we were unable to include chapters on Brazil, Portuguese Africa and Italian Ethiopia. In this sense we were very lucky in managing to retain a large selection of case studies and are grateful for the chapters that were submitted.

Beginnings: Hybrid Food Cultures and Foodways

The second chapter in the volume sets out to explore the very origins of settler colonialism in South America. Lisa Markowitz's 'Spanish Settlers and Andean Food Systems' describes the sixteenth- and seventeenth-century antecedents of the contemporary booming Peruvian foodscape. In the century before the Spanish invasion, the Andes were home to one of the great Native American empires, the Inka, which suffered the horrors of Spanish settler colonialism. The extractive appropriation of land and labour, however, did not result in the extinction of indigenous foodways. Rather, Markowitz shows that new hybrid foodways set root, predicated on the interpenetration of colonial and indigenous culinary economies. In the process local food mitigated pervasive dualisms: the separate spheres of Spanish/Indigenous, Coast/Highland and wheat/maize. In this rearranged world, indigenous foodstuff such as avocados, maize and passion fruit, as well as guinea pig (*cuy*), produced new culinary repertoires which attested to the ongoing exchange between coloniser and colonised. Throughout the centuries, Peruvian gastronomic hybridities embodied, mutated and reshaped the colonial encounter between indigenous peoples, enslaved Africans, Spaniards and the various combinations of their offspring, in public spaces such as workshops and markets, but also in private spaces sheltered by the intimacy of household and family. At the same time settler products such as the pisco sour became emblems of national foodways, celebrated as Peruvian authentic heritage.

The analysis of gastropolitics and culinary hybridity, as essential features of the original colonial encounter, are also present in the second chapter. In 'What Belongs in the "Federal Diet"? Depictions of a National Cuisine in the Early American Republic', Peter Mabli follows and examines the process of selection and exclusion of food items and recipes. Transitioning from colonial to a postcolonial and national polity, early American settler diet was a complex affair of rejecting certain Native American, European and African ingredients, while embracing others. Gradually, however, by the end of the eighteenth century, American political and social structures had evolved into entities distinct from their European colonial counterparts and settler food reflected a growing appreciation for the plentiful nature of the new land of milk and honey. The seemingly arbitrary choice to include some native items, such as maple syrup, yet exclude others (yaupon tea), in what may constitute a 'federal diet', defined a singular American cuisine largely affected by negative perceptions of certain cultures and peoples. Indeed, Mabli concludes, the identities formed using food were often deductive: settlers focused more on what they were not (not British, not Native American, etc.) more so than what they actually had become.

Across the Pacific in Taiwan, yet another complex variation of settler colonial food syncretism has unfolded over the past 500 years. In the chapter 'The Taste of Colonialism? Changing Norms of Rice Production and Consumption in Modern Taiwan', Yujen Chen takes an actual process of hybridisation—the development of new rice varieties on the island—as a means to explore the dynamics of native and settler colonial culinary identities in a context where the major colonial power (Japan) was not the main source of settler populations. Indigenous 'upland rice' (a variant of the Javanica breed) was, together with other local staples such as millet and taro, replaced from the seventeenth century with other rice varieties by the Qing masters of Formosa. Aboriginal peoples were gradually outnumbered by Han settlers and forced through land transfers, dispossessions and Imperial Chinese edicts to radically reorient the island's rice cultivation to the needs and tastes of the settler colonial and later immigrant populations. Formal annexation by Japan in 1895 further accelerated this process, imposing metropolitan preferences for short/medium grain Japonica rice—a trajectory that was paradoxically only reinforced as Taiwan became a new republic, home to mainland Kuomintang exiles. As Chen vividly illustrates in her chapter, intricate ecological, technological and geopolitical factors have over the past century combined with the continued contestation of rice flavour, texture and aroma within Taiwan, creating a market for idiosyncratic rice varieties like the now highly prized versions of 'Taiwanese Koshihikari'. The unquestionable transformation of Taiwan's agricultural systems and food cultures as a result of foreign occupation, settlement and immigration has however also witnessed the emergence of distinctly 'native Taiwanese' rice varieties and their accompanying dishes—many of which invoke aboriginal tastes in, for instance, taro-like aromas.

Similar processes of hybridity occurred in Mexico, which shares several parallels with the Peruvian case study. Despite (or perhaps because of) the violent dispossession and marginalisation of indigenous populations in Mexico, a distinctively 'republican' food repertoire emerged by the start of the twentieth century—almost a hundred years after national independence—where autochthonous staples like maize, chillies and squashes combined with imported ingredients and, in particular, invasive ungulates, to transform the country's culinary landscape. In a globalising context where nations are interpellated to define 'their' national or native cuisine, Mexican authorities in 2010 turned to the UNESCO's 'intangible heritage of humankind' by way of codifying a 'traditional Mexican cuisine'. In their chapter on 'mayoras', or traditional female cooks, Prieto Piastro and Colás tease out the tensions and contradictions between indigenous identities, market forces and institutional drivers negotiated by traditional cooks. At once purveyors of ancestral culinary cultures and gastronomic entrepreneurs (often dependent on public funding) 'mayoras' tend to find themselves recreating an idealised and largely (re)invented canon of traditional Mexican dishes, many of which employ postcolonial ingredients or techniques (the use of lard for cooking mole or distillation in the production of mezcal are cases in point). In the end, their contribution suggests, the process of certifying traditional Mexican cooks veers towards a liberal multiculturalism where the socio-economic, political and cultural hierarchies accompanying *mestizaje* are elided in favour of an atomised and essentialised performance of indigenous authenticity mainly tailored to heritage tourism and gastrodiplomacy.

From Erasure to Decolonisation

The culinary complexity of the colonial encounter, which blends rejection and expropriation with appropriation and incorporation, is illustrated in Adele Wessell's 'Unsettling the History of Macadamia Nuts in Northern New South Wales'. The Macadamia, an Australian native tree nut crop, reveals an entangled history of changing attitudes to indigenous flora and yet continuities between contemporary food culture and colonial attitudes. Originally eaten as a 'bush nut' it was commercialised in the nineteenth century, disrupting conventional representations of colonial distaste for native foods as simplistic and misleading. And yet, from the 1920s until 1997, it was rather Hawai'i that was the largest producer of the nuts in the world only to be temporarily replaced by Australia and currently by South Africa. The globalisation and commoditisation of macadamia nuts link large-scale economies to local histories of settler colonialism and dispossession. Food production was the motivation for settling the North Coast of New South Wales and used to rationalise the dispossession of Bundjalung Aboriginal people. In many ways, current agro-capitalism continues this history of disenfranchisement.

While Hawai'i was central to the globalisation of macadamia nut production, this by no means entailed a recognition of the Island's own indigenous heritage. Laura Kitchings' chapter entitled 'Definitions of Hawaiian Food: Evidence of Settler Colonialism in Selected Cookbooks from the Hawaiian Islands (1896–2021)' provides a critical account of such erasure. Analysing five cookbooks published over more than a century, Kitchings examines the unfolding definitions of Hawaiian foodways. The first two cookbooks reproduce a condescending settler approach by folklorising Native Hawaiian traditions and treating the Hawaiian feast, also identified as an 'Ahaaina', or 'Luau', as a large picnic. The third cookbook (1985) shows how the importation of a plantation labour force to serve the needs of the original settler colonialists led to a redefinition of Hawaiian foodways, one that embraced the diverse food cultures of the Islands but without specifically recognising Native Hawaiian food culture. In the twenty-first century, however, the narrative took a critical turn and the final two cookbooks examined demonstrate attempts to decolonialise the Islands foodways and acknowledge the history of settler colonialism in Hawai'i.

The contested definition of local food (settler and indigenous alike) and the challenge of decolonising gastronomy are the major themes in the heated debate about food politics and culinary appropriation in Israel/Palestine. Ronald Ranta and Daniel Monterescu provide an ethnographic account of these debates by framing the Jewish-Israeli foodscape as a struggle over definitions of authenticity and indigeneity. The chapter 'Decolonising Israeli food? Between Culinary Appropriation and Recognition in Israel/Palestine' examines whether the field of Israeli food is undergoing a paradigmatic shift with regard to the role and place of Arab and Palestinian people and food cultures. And, what such a shift might mean for Israeli settler colonial identity, society and attempts at decolonisation more broadly. Based on fieldwork and interviews, the authors identify three particular discourses adopted by Jewish-Israeli chefs with regard to Arab and Palestinian food: cosmopolitan appropriation, Arab-Jewish revival and reflexive decolonisation. While some chefs commodify Palestinian food as part of an urban hipster scene of consumption and others celebrate the return of Arab-Jews to their culinary source in the Middle East, a fledging yet marginal discourse seeks to recognise the prospects of decolonising settler colonial food from a reflexive and binational perspective.

The challenge of decolonising settler food culture is also evident in Jacqueline Newling's chapter '"Uneasy Lies the Head That Wears the Crown": Lamb or Kangaroo, Which Should Reign Supreme? The Implications of Heroising a Settler Colonial Food Icon as National Identity'. Drawing on historical references, period cookery texts and current scholarship—and using the emergence of lamb, rather than an indigenous animal, such as the kangaroo, as the 'national celebratory meat' on Australia Day, as her framework—Newling examines Australia's settler colonial heritage, food culture and foodways and its changing multicultural diversity and sensibilities. The chapter explores Australia's complex and contentious relationship with native food, which includes examples of erasure and appropriation, and highlights the uncomfortable truths of the dispossession of the country's first peoples. Her analysis provides an honest and reflective account of the debates surrounding the meaning and nature of decolonising Australian food and society more broadly.

After Decolonisation?

Two final chapters, on Algerian Pied-Noirs and white South Africans, respectively, attend most directly to one of the three core concerns of this volume: the phenomenon of 'self-indigenisation', or how 'settlers transform themselves into the native population, at least in their own eyes'. Inscribed in their self-description as *Afrikaaners*, the descendants of *Boer* settlers have, as Nitzan Tal explains in her chapter, no metropole to 'return' to once white supremacy was formally outlawed under democratic South Africa. This contrasts to the possibilities on offer to Anglo-settlers and other more recent white immigrants, who adopt a 'paradoxical identity: self-styled Europeanness on the one hand, self-styled nativeness on the other'. In a close and compelling reading of the 2010 novel *Double Negative* by South African author Ivan Vladislavić, Tal brings alive the power of food in settler colonial societies—its procurement, preparation, consumption and disposal—through material objects like barbeques, ramekins, cafetières and discarded shopping lists. In the novel, the end of South African Apartheid upends some of the country's previous alimentary markers of distinction—the *braai* as a symbol of frontier masculinity; the soufflé of an effete Anglo urbanity; the coffee pot, of cosmopolitan worldliness; mealie meal, the racialised staple of non-whites—but then throws them back at a supposedly democratised society in the form of contemporary culinary repertoires that seek to 'renegotiate both Europeanness and locality' without, it seems, being able to shake off the enduring structural power of race and class.

While whites in the new South Africa seek to reaffirm their nativeness through foodways that reinvent locality and worldliness, Algerian Pied-Noir communities now exiled in France use memory as the vehicle to reproduce settler colonial food identities in their adopted homeland. Here, 'self-indigenisation' is a process accompanying decolonisation, but one which unfolds in a distant past, *away* from the native land, not *within* the present post-colony, as is the case for many South African whites. 'In exile, belonging to and coming from Algeria became central to Pied-Noir identity', Amy Hubbell and Jorien van Beukering suggest in their chapter entitled 'Sustaining the Memory of Colonial Algeria Through Food', 'and their memories strove to authenticate this "indigenous" identity. … In other words, the Pieds-Noirs cultivated attachments to Algeria as their identity became linked to a country that was now absent'. Through their comprehensive analysis of Pied-Noir cookbooks and contemporary food websites, Hubbell and van Beukering reveal several of the core themes animating this volume: the construction of 'white native' identities (in this case, Pied-Noirs) through the cultural homogenisation of otherwise disparate European settlers, migrants and refugees, all at the expense of the rights of existing native populations; the appropriation of native food items (merguez, couscous, méchoui) for the (re)colonial culinary repertoire; and the radical reconfiguration of land use, ownership and access (including the revalorisation of metropolitan *lieux de mèmoire* such as the annual Catholic Ascension Day pilgrimage to Notre-Dame de

Santa-Cruz in Nîmes-Courbessac). This chapter, however, also draws attention to different forms of settler colonial memorialisation through food. Some first-generation Pied-Noirs indulge in a 'restorative' nostalgia where the recreation of 'authentic' recipes from 'la-bas' ('back there', in Algeria) brings alive in France a single, romanticised view of the past, tinged with a sense of loss and victimhood (a 'nostalgérie'). This contrasts to what Hubbell and van Beukering identify as a 'reflective' nostalgia more common among second- or third-generation Pied-Noir descendants, who tend to embrace a more diverse, flexible and productive reading of the Pied-Noir culinary legacy.

The chapters in the volume raise a number of pertinent questions regarding the decolonisation of settler colonial foodways and food cultures and their relationship with indigenous people. Reflecting on these questions, Lorenzo Veracini argues in the volume's final chapter, 'the predicaments of settler gastrocolonialism', the need to put indigenous people and their right to food sovereignty at the forefront of any discussions on decolonisation. Decolonisation, according to Veracini, needs to start from the acknowledgement 'in a non-tokenistic way' of the historical links and rights of indigenous people to their land. The commodification and use of indigenous food, including its 'harvesting, production, preparation, and consumption', must directly benefit indigenous communities. It is evident, reflecting on the volume's case studies, that real decolonisation can only occur 'when the Indigenous collective is able to self-determine its modes of social reproduction, a passage that must include a full recovery of its food sovereignty'.

Bibliography

Abufarha, Nasser. 2008. Land of Symbols: Cactus, Poppies, Orange and Olive Trees in Palestine, Identities. *Global Studies in Culture and Power* 15 (1): 343–368.

Allemann, Samantha. 2018. Indigenous Foods: Swallowing Our History. *Matters Journal* (May 2018): np. Accessed October 16, 2021. https://www.samantha-allemann.com/food/swallowing-our-history-matters-journal/.

Banerjee-Dube, Ishita, ed. 2016. *Cooking Cultures: Convergent Histories of Food and Feeling*. Cambridge: Cambridge University Press.

Bardenstein, Carol. 1998. Threads of memory and Discourses of Rootedness: Of Trees, Oranges and the Prickly-Pear Cactus in Israel/Palestine. *Edebiyat* 8: 1–36.

Bell, David, and Gill Valentine. 1997. *Consuming Geographies: We Are Where We Eat*. New York: Routledge.

Bhabha, Homi K. 1997. *The Location of Culture*. London: Routledge.

Comaroff, John L., and Jean Comaroff. 2009. *Ethnicity, Inc*. Chicago: University of Chicago Press.

Daigle, Michelle. 2019. Tracing the Terrain of Indigenous Food Sovereignties. *The Journal of Peasant Studies* 46 (2): 297–315.

Fanon, Franz. 1991. *Black Skin, White Masks*. London: Pluto Press.

Gabaccia, Donna R. 1998. *We Are What We Eat: Ethnic Food and the Making of Americans*. Cambridge, MA: Harvard University Press.

Grey, Sam, and Lenore Newman. 2018. Beyond Culinary Colonialism: Indigenous Food Sovereignty, Liberal Multiculturalism, and the Control of Gastronomic Capital. *Agriculture and Human values* 35: 717–730.

Harris, Jessica B. 2012. *High on the Hog: A Culinary Journey from Africa to America*. New York: Bloomsbury.

Hixson, Walter L. 2013. *American Settler Colonialism: A History*. New York: Palgrave Macmillan.

Mamdani, Mahmood. 1998. When Does a Settler Become a Native? Reflections on the Colonial Roots of Citizenship in Equatorial and South Africa. Inaugural Lecture, University of Cape Town, New Series no. 208.

Memmi, Albert. 1985. *Portrait du colonisé, précédé de portrait du colonisateur [The Colonizer and the Colonized]*. Paris: Gallimard.

Mintz, Sidney. 1985. *Sweetness and Power: The Place of Sugar in Modern History*. New York: Penguin Books.

Mintz, Sidney W., and Christine M. Du Bois. 2003. The Anthropology of Food and Eating. *Annual Review of Anthropology* 31 (1): 99–119.

Monterescu, Daniel, and Ariel Handel. 2019. Liquid Indigeneity: Wine, Science, and Colonial Politics in Israel/Palestine. *American Ethnologist* 46: 313–327.

Ranta, Ronald, and Atsuko Ichijo. 2016. *Food, National Identity and Nationalism: From Everyday to Global Politics*. Basingstoke: Palgrave Macmillan.

Ranta, Ronald, and Yonatan Mendel. 2016. *From the Arab Other to the Israeli Self: Palestinian Culture in the Making of Israeli National Identity*. London, Routledge

Veracini, Lorenzo. 2010. *Settler Colonialism: A Theoretical Overview*. Basingstoke: Palgrave Macmillan.

———. 2011. Introducing Settler Colonial Studies. *Settler Colonial Studies* 1 (1): 1–12.

———. 2013. The Other Shift: Settler Colonialism, Israel and the Occupation. *Journal of Palestine Studies* 42 (2): 26–42.

———. 2015. What can Settler Colonial Studies Offer to an Interpretation of the Conflict in Israel-Palestine? *Settler Colonial Studies* 5 (3): 268–271.

Wilk, Richard. 2006. *Home Cooking in the Global Village: Caribbean Food from Buccaneers to Ecotourists*. New York: Berg.

Wolfe, Patrick. 1999. *Settler Colonialism and the Transformation of Anthropology: The Politics and Poetics of an Ethnographic Event*. London: Cassell.

———. 2006. Settler Colonialism and the Elimination of the Native. *Journal of Genocide Research* 8 (4): 387–409.

Part I

Beginnings: Hybrid Food Cultures and Foodways

2

Spanish Settlers and Andean Food Systems

Lisa Markowitz

Introduction

Peruvian food has been having a long moment. For the past two decades, the country has enjoyed a gastronomic boom. It is centred in Lima, where restaurants have drawn Michelin stars and legions of customers for elegant reinventions of classic dishes and for introducing and integrating ingredients and preparation styles from the Andean nation's multiple agro-ecological zones, immigrant populations, and regional foodways (Álvarez 2007). Peruvian dishes have reaped fans across the globe, while Lima has garnered international awards and chefs have enjoyed transnational celebrity and commercial success. The promotion of gastronomy is a broad-based project shared by government agencies, celebrity chefs, food writers, nongovernmental organizations, and commercial associations. Indeed, the work of fostering a greater appreciation of the country's variegated cuisines and foods has assumed a political cast, as a means

L. Markowitz (✉)
Department of Anthropology, University of Louisville, Louisville, KY, USA
e-mail: lisam@louisville.edu

R. Ranta et al. (eds.), *'Going Native?'*, Food and Identity in a Globalising World,
https://doi.org/10.1007/978-3-030-96268-5_2

both to stimulate economic prosperity and to nourish national identity. This former aim has met some sectoral successes, at least pre-pandemic, with a swelling restaurant sector and tourism industry (e.g. Contreras 2011; Khalip 2007; McLaughlin 2011; Ortiz 2010; and Rovzar 2016). The latter, a high-profile campaign to conjoin cuisine with national identity, raises questions about the very nature and predicates of this ostensibly inclusive agenda, drawing scepticism about such matters as substantive participation in constructing "national food" (e.g. Matta 2016; Passidomo 2017; Takenaka 2019), the relative material benefits of this project (e.g. Kollenda 2019; López-Canales 2019; Markowitz 2018), and the ways it reprises colonial frames of power and extraction (e.g. Garcia 2013, 2021; Grey and Newman 2018; McDonell 2019). In this chapter, I take a step back from the ongoing appraisal and critique of contemporary gastropolitics to describe their sixteenth- and seventeenth-century antecedents. Just as trajectories of conquest and settlement shaped Peruvian diets, they generated enduring exclusions and antagonisms that rive Peruvian society today. These are legible in foodways.

Settler Colonialism

Let me begin with the beverage most associated with Peruvian dining, the pisco sour. The cocktail (grape brandy, lime, egg whites, simple syrup, with a touch of Angostura bitters) is served in restaurants and homes across the country's multifarious physical and cultural landscapes. An emblem of national foodways, the pisco sour, like its U.S. counterpart apple pie (Patel 2021), signals a grim history beneath its frothy surface. Colonial legacies, however deliciously transformed in the present, adhere to the grape. As historian Rebecca Earle (2012) has recounted, the Spaniards who entered the western hemisphere at the end of the fifteenth century to conquer, plunder, and settle were enormously concerned with eating properly. Since one of their priorities was procuring decent wine, the Spanish worried about the agronomic potential for grape cultivation across the diverse locales of their newly claimed territories (Crosby 2003). It was in the river valleys of southwestern Peru that productive viticulture began to emerge in the 1550s (Davis 1984, 48–49), scarcely two decades after the capture of the Inka Atahualpa, an event marking the beginning of the end for the Inka Empire. Over the next 100 years, the regional wine industry expanded, fuelled by demand from one of the economic hubs of Spanish America: the silver mines of Potosí. Site of one of the richest strikes in history, Potosí, by the early seventeenth century, was one of the world's largest cities. On the high plain of what is today the southern Bolivian *altiplano*, impressed and dispossessed native labourers hauled rock up out of mineshafts and crushed ore in perilous conditions. Stein and Stein (1970) have made the case that the treasure of Potosí, funnelled through the Spanish elite to northern European merchants and manufacturers, served as seed capital for the emergence of capitalism.

Sugar, another ingredient, claims a bitter globalized legacy as well. "The demographic disaster, provoked by the sixteenth century conquest of Peru that caused the collapse of the native population, opened the way for the importation of African slaves and the rise of slavery, particularly in the cultivation of sugar on the coast," notes Peter Klaren (2005, 34) in an economic history of the crop. Spanish settlers on the dry north coast of Peru established *haciendas* to supply basic foods to nearby towns. Sugar, in contrast to the other staple items, required hefty capital investment in technology and labour. Epidemic diseases and the forced removal and resettlement of indigenous residents (*reducciones*) meant growers could not satisfy their labour needs with local populations and had to look elsewhere. With outright enslavement of any remaining native people forbidden by the Spanish Crown, the landowners turned to importing enslaved Africans. Through the seventeenth century, as sugar prices rose and sugar plantations covered large swathes of the northern and central Peruvian coastline, the number of slaves rose two- to threefold. As late as 1821, about 40 per cent of the enslaved people in Peru worked in sugar (Klaren 2005, 35–36).

I could continue the deconstruction of the pisco sour to explore other dismal food chain linkages (eggs and the recent appearance of Peruvian poultry Confined animal feeding operation (CAFOs) come to mind). The point here is that the examination of contemporary foodways reveals deeper histories and connections. In this chapter, I highlight the early moments of the colonial encounter to trace the reconfiguration of Peruvian agriculture and diet, emphasizing the impacts of the introduction of European animals and cultigens on native production systems. The nearly 500 years between the Spanish invasion of the lands that would become the modern state of Peru and the current discursive celebration of Peruvian food direct attention to the complexity of historical processes and the corresponding challenge of distilling centuries of biological and cultural imperialism into a chapter-sized précis.

As the economic and political dynamics described in these introductory vignettes suggest, food also serves as a way into a wider discussion of settler colonialism in Latin America. Lucy Taylor (2020) observes that Latin American countries, in general, have been outside the core of settler colonial scholarship with its predominant focus on North America, Australia, and Israel-Palestine. She notes that attention to Latin America offers an opening to deepen and further enrich settler colonial theory via consideration of capitalism, global hierarchies, Afro-descendants, and assimilation. My emphasis here is on the section of her analysis labelled capitalism which, for reasons of historic periodization in the case of Peru, might be more aptly called exploitation. Peru, like Taylor's subject, Argentina, complicates the "rule of thumb" distinction between appropriating land and appropriating labour (Taylor 2020, 5). Simply put, the Spanish did both. Shannon Speed (2017, 785) makes a similar point about the region, "where in many countries 'Red people's' labour was applied to 'Red people's' land, leaving indigenous peoples subject to *both* territorial dispossession and bodily exploitation." Speed goes on to underline the relevance to Latin America of Patrick Wolfe's premise that "the colonizers came to stay," making "invasion…a structure not an event." In Latin America, as in other settler societies, woven through the structures is ingrained, if little voiced, fear of the numerically greater underclass (Gott 2007).

In Peru, forms of domination and exploitation engendered by early Spanish organization of their new territories and control over their new "subjects" persist. Some of these forms are institutional. For example, *haciendas*, established in the sixteenth century and proliferating through the Republican era, maintained essentially feudal production relations, which lingered into the 1960s before they were officially declared illegal as part of the country's Agrarian Reform (1968–73). Anthropologist Paul Doughty (2001, 239) notes that in 1961, estates with "resident serf populations…constituted 80% of all agricultural land."[1] Some of the continuities take a more amorphous if deeply experienced form in the racialized inequalities affecting people of indigenous background and rural provenance. Although for much of the past two decades Peru has had one of the world's fastest growing economies, thanks largely to mineral exports, and nationwide poverty rates fell by more than half, rural poverty rates are still three times those of urban areas, and living conditions in the Andean regions remain far below those on the coast (Nava 2015, 5).[2] A speaker of a native language such as Quechua or Aymara is twice as likely to be poor as a Peruvian whose first language is Spanish (ibid.: 6). As I have argued elsewhere (Markowitz 2018), these inequalities have reflected and reified a national food hierarchy, one privileging European over indigenous items. If the structures of invasion have enforced multifold forms of disparagement of indigenous foodways, how were these structures set in place by the events of Spanish settlement? How in particular did the settler appropriation of land and labour affect production and consumption practices? In the following sections, I address these questions by sketching native food production and diet and some of the biological and economic dimensions of Spanish settlement.

Indigenous Foodways

In the century before the Spanish invasion, the Andes were home to one of the great Native American empires, the Inka, and, well before their rise in the fifteenth century, to a series of sophisticated, complex societies. Inka elites and administrators oversaw a dual food system: household and community-based subsistence production and a state-orchestrated agricultural regime concerned with provisioning the Imperial specialists (priests, craftspeople, and soldiers), ensuring supplies of ceremonial and

[1] In the southern department of Puno, reconstitution of holdings expropriated during the agrarian reform is occurring and landowners I interviewed in 2014 and 2016 referred to their holdings as *haciendas*.

[2] COVID-19 has exposed the structural fragilities of the economic success story, notably underinvestment in health care and the high rates of labour informality (Taj and Kurmanaev 2020). See also the 2018 IPSOS survey for data on people's experience of prejudice (Ipsos 2020).

sumptuary goods, and rewarding and supporting their far-flung, multi-ethnic subjects. The land under Inka control stretched along the western coast of South America, from today's Santiago to Quito, incorporating about a million square kilometres. The high altitudes, glaciers, and inter-mountain valleys of the world's longest mountain range, as well as the particular climatic characteristics of tropical mountains and the influence of the Pacific Ocean, create a complex physical environment. This same complexity gave rise to the domestication of numerous crop species. The opening line of *Lost Crops of the Incas* reads, "At the time of the Spanish conquest, the Incas cultivated almost as many species of plants as the farmers of all Asia or Europe" (NRC 1989, 1). The Inka-controlled lands coincident with Peru can be divided into three distinct regions: a dry strip of coastline cut by river valleys; the Highlands, a patchwork of valleys, steppes, ridges, and peaks; and the eastern escarpment and lowlands—each region affording and constraining productive possibilities. D'Altroy (2015, 99) underscores this socio-ecological complexity, noting that "the variegated landscape created a myriad of different micro-environments to which societies had to adapt their foraging, herding, fishing, and agricultural strategies."

The arid western slopes of the Andean range are ribbed by a series of rivers flowing to the Pacific. The intensively irrigated river valleys supported the cultivation of cotton, gourds, corn, squash, and beans. Valley residents enjoyed relative proximity to the riches of the littoral itself. The cold Humboldt Current flows north along the dry Chilean and Peruvian coastlines, supporting a great variety of marine life, notably small bony fish, like anchovies and sardines, which in turn sustain much larger species. During Inka times, thousands of specialized fishing communities were scattered along the coast (Marcus et al. 1999). Excavations from Cerro Azul, a fishing settlement 130 kilometres south of Lima, offer a glimpse of the maritime diet. Using cotton nets, fishers stood near the beach itself on rocky ledges over the sea or paddled close to shore, to capture a variety of species. The local catch included sardines, anchovies, and 20 types of larger fish, among them mullet, corvina, flounder, and catfish. Residents also seem to have supplemented their diet with inland products, likely trading dried anchovies and sardines for fresh and jerked llama meat.

Heading upslope and east, we come to the Inkan heartland, the Andean Highlands, where about 60 per cent of the indigenous population lived at altitudes of 3100 metres or above (D'Altroy 2015). At the lower elevations, farmers raised beans, vegetables, quinoa, tubers, and, most importantly, maize, the sacred *sara*. Krogel (2011) notes that the large number of Quechua words used to denote the plant varieties and preparations point to its cultural centrality. She also observes that the Spanish-language chroniclers seemed to emphasize the difficulty and tedious labour of cultivating this plant, perhaps to indicate the high culinary and ritual estimation in which it was held. While maize was grown across Andean communities, its large-scale production, involving investments in extensive irrigation and terracing, was a state undertaking directed towards supplying the military and religious and ceremonial use (Murra 1980). Cornmeal also furnished the major ingredient for *chicha*, a fermented corn beer. Above 3500 metres, where frosts precluded maize cultivation, farmers turned to crops more suited to the cold and a shorter growing season. Legumes included the large-seeded lima beans (which take their name from Peru's coastal capital) and the lysine-rich *tarwi* (NRC 1989, 181). Highly nutritious Andean grains—or, more accurately, chenopods—are also common at these altitudes. In addition to the current favourite of the gluten-adverse, quinoa, Highland people cultivated *kiwicha*, a kind of amaranth, and the especially hardy *kaniwa*. The sixteenth-century chronicler, the Inca Garcilaso de la Vega, describes the consumption of quinoa: "The Indians and Spaniards eat the tender leaves in their braises because they are tasty and healthy; they also eat the grain in their vegetable stews, prepared in many ways. The Indians prepare a brew from the quinua to drink, like one made from maize, but made in lands where there is a shortage of maize. The Indian herbalists use quinua flour for some sicknesses" (cited in Krogel 2011, 29–30). Of all the Highland domesticates, potatoes stand out as the most important staple; indeed, today potatoes are the world's third most important food crop, after wheat and rice. Andean farmers have domesticated some 3000 varieties of potatoes (Peruvian Ministry of Agriculture 2008), selecting for different culinary and agronomic qualities—resistance to drought, frost, and pests. Indigenous potatoes vary in appearance as well, coming in many different shapes and sizes, their flesh an array of violet, white, red, and deep yellow and various combinations of these. In contrast to maize, potatoes were much more of a commoner's subsistence-oriented crop (Murra 1980). Highland people also domesticated "lesser" tubers, unknown beyond the Andes, including the hardy *oca* and the brightly coloured and delicious ulluco or *papa lisa*.

At elevations above 4000 metres, conditions preclude most cropping, save for some tubers. Here on the cold, rolling expanses of prairies, Andean shepherds tended herds of native camelids: alpacas and llamas. Although alpacas are known for their fine fleece and association with brilliant weaving traditions (and, today, high-end sweaters), llamas served for portage, both also featured on the Inkan menu. It is important to keep in mind that in the Americas very few animals were domesticated, and none of these species could come close to llamas or alpacas in furnishing as much dietary protein per carcass. Archaeological analyses indicate that ensuring a regular supply of meat would have been the primary goal for camelid domestication (Aldenderfer 2001; Moore 2016). At the time of the Spanish invasion, the distribution of camelids was much wider than it is today (Flores Ochoa 1988) and the consumption of their meat was probably common in some form (fresh or jerked) in *altiplano* homes (Murra 1980, 49). In addition, llamas played a crucial role as pack animals enabling the exchange of food and other valued items between different resource zones.[3]

To conclude this overview of Andean productive zones, we drop down onto the eastern slopes as they descend into the Amazon rainforest. Here again, the ecozones have different productive capacities at different elevations. The upper edges of the rainforest permitted maize, and then at lower altitudes, coca, a huge variety of fruits, and other warm-weather crops, while the forest resources included wood and feathers (D'Altroy 2015). Although the gastronomic boom has generated fresh interest in and demand for foods from the Andean Amazon, we have relatively few records of native foodways in this region.

---

[3] Llama pack trains hauled foodstuffs between different environmental zones up through the twentieth century although this diminished with the expansion of road construction and increased frequency of truck transport. In a practice known as *trueque*, caravaneers would depart their homes in high, pastoral zones to barter set measures of meat for those of corn or other lower-elevation products and then travel on to trade the acquired items for other goods.

A final theme in this sketch of indigenous agro-foodways is preservation itself. At a macro level, the Inkas connected their disparate multiethnic population through a road system and, along those roads, built a vast state-maintained system of food storage, stocking storehouses with supplies, including a large variety of foodstuffs for their armies, for state functionaries, for recourse during lean years, and for redistribution among the inhabitants of their empire (D'Altroy 2015; Krogel 2011; Nair and Protzen 2015). They designed the warehouses to protect the food against deterioration from the sun, wind, and humidity (Krogel 2011, 60). The techniques of preservation included the strategic use of environmental zonation. Rostworowski (2007, 127) explains that the Incas stored maize kernels in large, covered earthenware jugs in high-altitude facilities where the cold would prolong storage and reduce threats from insects and rodents. At the household level, people used the intense climate to preserve foods. Freeze-drying potatoes to prepare *chuño* was and is ubiquitous in the Highlands. While recipes vary across the region, in brief, people take advantage of the many frosty nights and intense sunshine of the Andean winter to freeze "bitter" potatoes (one of the myriad potato varieties) overnight and then during the day, expel the moisture and leave them to dry in the sun. The desiccation process takes several days and results, when the potatoes are rehydrated, in a dense and filling food, used in soups and stews and as part of a portable cold lunch. *Chuño* keeps for years. The frigid Andean nights also permitted the preservation of llama and alpaca flesh as *ch'arki*, the origin of the word jerky. Thin strips of meat would be hung outside for weeks, left to freeze overnight, and then to dry during the day. These practices all speak to a close knowledge of ecological possibilities as well as the need to reserve food supplies against poor harvests and other disruptions. However, the imminent European arrival would bring about social upheaval no one could have imagined.

The Settlers

Settler colonialism is associated with the elimination of the native population, a practice that has taken multiple forms across time and space (Wolfe 2006). In the case of Andean America, the Spaniards did not aim to eliminate native *people* but rather to use their labour on their newly annexed territories. The conquistadores, however, did not arrive in the western hemisphere alone; the colonial enterprise was imbricated in a massive transfer of plants, animals, and microbes. In his pointedly titled *Born to Die: Disease and New World Conquest, 1493–1650*, Cook (1998) recounts the horror of the sweeping epidemics. New World peoples had no resistance to Old World pathogens, and infectious diseases—airborne smallpox, measles, and influenza being the most lethal—ravaged the native populations. The epidemics came in waves, bringing excruciating death tolls and terror. Widespread illness meant that households and kin groups could not adequately care for the sick and young or tend to animals and crops, exacerbating grim conditions with worsening nutritional health and despondence. Newson (1993) calls the collapse of New World native populations—evidence suggests a fall of 90 per cent between 1492 and 1650 (277)—a demographic disaster with "no known parallel in world history" (253). Differences in mortality rates over space and time point up not only the variable regional impacts of disease but the different forms of violence associated with invasion and settlement, and their interactions (Livi-Bacci 2006), as well as the methods of control and exploitation the Europeans employed (Newson 1993).

In less cataclysmic ways, the arrival of Old World cultigens and domesticates[4] induced widespread suffering. The Spanish viewed food as a matter of commerce *and* of person. "The successful exploitation of the New World by these people depended on their ability to 'Europeanize' the flora and fauna of the New World. That transformation was well underway by 1500, and it was irrevocable in both North and South America by

[4] Janer (2007, 389) points out the asymmetry of this cross-hemispheric transfer, with the Americas contributing a much greater number of items to the global pantry than the Europeans. The scale of the exchange sometimes startles, not only in regard to biota but in respect to conventional assumptions about primordial foodways, for examples, Italians and tomatoes, Irish and potatoes, or to iconic agro-industries, for instance, cowboys and gauchos, banana republics.

1550" (Crosby 2003, 64). The transplanting of Iberian agriculture and foodways underlay the very survival and commercial success of the colonies, permitting both their rapid expansion and the ability to support the crucial mining sector (Super 1988). For the colonists themselves, having familiar foods was essential to bodily and spiritual well-being (Earle 2012). While the Spaniards grew to appreciate and enjoy certain New World foods, such as chocolate, pineapples, chillies, and sweet potatoes, it was their own staples—wheat bread, beef, pork, and wine—that were considered the mainstays of a diet appropriate for a European. For Spaniards occupying the former Inka lands, this implied the need to establish rights to land and the ability to muster labour to plant those essential foods and then cultivate them on a large enough scale to supply other settlers and ultimately the economic hubs of Lima and Potosí. Occupation of land and exploitation of labour took multiple and shifting forms in Spanish America. Here, I emphasize the institutions and practices most associated with the initial transformation of the food system.

Land and Labour

In the 1530s, the Governor of Peru, Francisco Pizarro, followed the common Spanish American practice of rewarding his lieutenants in the invasion, as well as other settlers of consequence, with a grant, known as an *encomienda*, to the labour of people living in a designated area. Created to operate as a form of indirect rule, the *encomienda* system predominated for a generation or so after the Spanish invasion. Other than to visit the area to determine appropriate tribute payments and presumably arrange for the care of native souls, grantees or *encomenderos* were meant to leave the Indians alone. They typically managed the land grant and people via an alliance of sorts with the local indigenous leader (*curaca*), and the small Spanish population had relatively little contact with the still-sizeable indigenous one (Morner 1985).[5]

[5] See Stern (1982) for an account of the perspectives of the *curacas* as they dealt with *encomenderos* in Huamanga (Ayacucho).

An example from southwestern Peru reveals how even indirect extraction transformed Andean agriculture (Davis 1984). Initially, *encomenderos* living in and around the newly founded town of Arequipa (today, Peru's second largest city) collected their tributes in the form of local products that had limited market resale value. Seeking more lucrative commercial possibilities, the grantees began to require payment in European crops and animals. By the mid-century, the growth of Lima and Potosí generated increased demand for these. We get a sense of how much the agricultural conversion was underway from a 1549 tribute list: 750 bushels of wheat, 15 pigs, and 50 sheep and goats, among other goods (Davis 1984, 23). Davis concludes by noting that in Arequipa as elsewhere in Peru, in the second part of the sixteenth century, tribute revenues declined. As the recurrent epidemics battered their communities, indigenous people found it impossible to keep up with tribute demands, often levied with little regard to their feasibility.

Providing the Spanish with desirable foodstuffs altered Andean landscapes at all elevations as it involved inserting European species into plots and pastures and replacing indigenous ones. Morner (1985, 46–47) describes the changes above the Urubamba-Vilconata River Valley, near Cuzco. On the lower hillsides, up to 3300 metres, a variety of Old World vegetables and fruits joined the still predominant maize for which the Spanish had developed a taste. Higher up, quinoa and potatoes had to make room for wheat, barley, and broad beans. And, in the high herding zones, sheep moved into alpaca pastures (Flores Ochoa 1988). The last of these speaks to a widespread dislocation of indigenous camelids. In the sixteenth century, European ruminants, afflicted with mange plagues, transmitted this contagious ailment to alpacas and llamas, decimating herds and destroying the basis of pastoral livelihoods.

In response to a suite of economic and political problems besetting the Andean region, a new emissary of the Spanish Crown, Viceroy Francisco de Toledo, initiated a reorganization of Indian management in the 1570s. Envisioning transformation of native settlement patterns, he implemented the creation of *reducciones* or resettlements for the indigenous population. No longer were people to reside in their kin-based communities known as *ayllus*, but rather were to relocate to European-style nucleated settlements. As Stern (1982, 77) observes, this model "conflicted sharply with traditionally dispersed settlement patterns designed to exploit a series of scattered Andean microenvironments." Concentrated populations might be more easily counted, taxed, converted to Christianity, and mustered for labour drafts to toil in Potosí and later in the mercury mines of Huancavelica. Uprooted and crowded together, people were also more vulnerable to epidemic disease. In what would become the Province of Caylloma in the southern Highlands, Toledo's plan involved the establishment of 24 settlements, each with a church and town hall set upon a central plaza. Caylloma affords an interesting case study since, unlike many of the engineered communities, all 24 villages remain today. (It is not without irony that in the 1990s the Spanish national development agency restored several of their churches as tourist attractions. Local residents grumbled about the Spaniard developers and the "reconquest.") In an exhaustive ethnohistoric and archaeological land-use analysis, Steve Wernke (2013) explicates the ways local *ayllus* negotiated with the Spanish to determine the "emplacement" of these communities. Even in light of the compromises, resettlement undermined food production. In the wake of the epidemics, the surviving population, faced with a diminished pool of workers and longer walks to their dispersed plots, ultimately abandoned irrigation systems and even large productive fields.

Concurrent and later Spanish extractive institutions—the labour draft, reformulated tribute demands, and *haciendas*—further disrupted indigenous agriculture and dismantled community structures. Dislocated indigenous survivors managed through commerce and labour to make their way into Spanish towns and adopted some European tastes, while the settlers and their descendants integrated New World foods into their diets.

Hybrid Foodways?

In this rearranged world, indigenous people encountered the new crops and animals they were obliged to tend along with less reliable access to their familiar foods. For their part, even as the Spaniards endeavoured to replicate their Iberian diet, they confronted myriad novelties and the need, particularly for those of the middling and lower classes, to make do with local substitutes. A great irony of the Columbian exchange, given its cost in human suffering, was in fact the quality of the *available* (not to be conflated with accessible) diet that resulted. John Super's 1988 economic history of food in sixteenth-century Spanish America anticipates current scholarly and popular interests. Focusing on circumstances in Mexico and the central Andes, he points out the potential for settlers, the enslaved, and natives to eat well or at least adequately. While Europe suffered periodic famine, Spanish America, in general, enjoyed good yields of grain, abundant animal protein, and fresh foods much of the year (Super 1988, 80–81). The combination of Old and New World foods allowed the possibility for "one of the best diets in the history of the world" (23).

For subjugated peoples, selecting dietary and productive options occurred in the context of settler coercion. Kubler (1946, 354–59) forwards criteria by which indigenous people might have evaluated the uptake of European species to generate the cash needed to pay tribute and perhaps provide some additional household benefits. Did the plant or animal offer qualities unmet by local ones? For example, equines and cattle provided traction and portage far surpassing those of the llama. Sugarcane, which flourished in deep Andean valleys, was widely relished as a sweetener while the stalks served as a fuel source. Next, a desirable new crop would not impose additional labour burdens (unless its value was great). Therefore, in contrast to sugar cultivation, the exacting routines of viticulture appealed little to indigenous farmers who preferred *chicha* or *aguardiente* (sugarcane whiskey). Third, in some cases, take-up had to do with links between crops: raising cattle and horses required forage crops like alfalfa. Finally, environmental considerations, especially the altitude, may have been paramount. While some of these are based on obvious agronomic or veterinary limitations, others are more social in the sense that, through the colonial period, the higher elevations became increasingly associated with indigenous residence and diets.[6]

[6] The disparagement of foods associated with high-altitude residence persisted through the twentieth century. For example, until recently most urban Peruvians considered the consumption of alpaca or llama unthinkable (Markowitz 2012).

Zooarchaeologist Susan deFrance (2021) develops this point about food as a geographic and social boundary marker in her look at colonial-era Spanish treatment of the guinea pig. Relatively easy to home-raise, *cuyes* were and continue to be eaten across the Andes, typically as a special occasion dish. Reviewing faunal collections from 11 diverse colonial sites in Peru and Bolivia, deFrance found that guinea pigs constituted a minuscule proportion, less than 1 per cent, of mammalian remains. Materials from the earliest sites suggest that guinea pig use was quickly supplanted by Old World animals (135). She contends that the Spaniards considered eating this common Andean meat animal anathema. Consuming native food jeopardized a Spaniard's bodily and spiritual well-being (Earle 2012), and *cuy*, perhaps because it featured in Andean ritual practices, which the Spanish viewed as idolatrous, was, more so than other indigenous species, culturally out of bounds. Spanish rejection of the meat may have had a role in actually motivating indigenous production and the concomitant association with native life, for *cuyes*, unlike other animals, were not subject to tribute extraction or seizure. Reinforcing the gap in respective dietary preference, deFrance suggests, was physical separation. The concentration of native peoples in the higher elevations came about through colonial relocation policies and the higher losses to epidemic disease suffered by coastal populations, while Europeans at altitude struggled with hypoxia and preferred to reside in lower valleys. Without losing sight of the role of dietary distinctions, another explanation for the limited appearance of guinea pigs is the proliferation of European livestock. Super (1988, 29) reports that during the sixteenth century meat production boomed throughout Spanish America and prices were low, particularly for pork and beef. He refers to native peoples as "great lovers of meat," noting that even as beef prices later rose, pigs, poultry, goats, and sheep remained available (56).

Spaniards also succeeded in provisioning themselves with another core element of their accustomed diet. "The absence of wheat in the Americas…was the epitome of the biological divide between Europe and the New World. If bread eaters were to settle in the Americas, wheat had to accompany them" (Millones Figueroa 2010, 304). Spanish respect for maize's qualities and their recognition of it as a kind of New World analogue grain did not deter them from planting wheat everywhere it would grow. Although the Andean region was not the most hospitable for it, wheat cultivation was well underway by the mid-sixteenth century (308). A few decades later, settlers in temperate valleys raised enough to supply bakers in Potosí and Lima; by 1603, 28 bakeries in Potosí could purchase flour sufficient for 36,000 lbs of bread a day (Super 1988, 35). We can surmise that the turnover of high-quality fields to the settler staple probably did not endear it to indigenous farmers.

What did Andean people think about the Spanish gastronomic juggernaut? Eighteenth-century painters may offer us an inkling (Krogel 2011, 64–67). The colonial Church had decided that visual art offered a good means to indoctrinate native peoples in the faith. Consequently, it supported the production of European-style images and the training of the image-makers. The Cusco School, painters influenced by the Baroque tradition, depicted biblical scenes including renderings of the Last Supper that feature, at centre, a guinea pig. Krogel analyses the Last Supper painted by the best known of the Cusco group, Marcos Zapata, pointing out his insertions of native foods—avocados, maize tamales, and passionfruit, as well as the guinea pig, into the sacred Christian feast, and suggests that he and the other painters were poking at the arrogance of Spanish assumptions about their own cultural superiority. Replacing the wheat loaf shared by Jesus and his apostles with small cornbreads (Millones Figueroa 2010), and the substitution of the reviled *cuy* for lamb, does suggest a dig at European dietary sensibilities.[7]

While cultural preferences, social and physical location, and most certainly economic exigencies promoted different diets, patterns of labour and commerce contributed to the development of shared tastes. Despite the desire of the Spanish Crown for Indians and settlers to occupy segregated administrative and residential spheres (a Republic of Indians and a Republic of Spaniards), the demands of the *encomienda* system and the *mita* (labour draft) meant that large native populations regularly worked in and around urban centres. *Someone* had to construct houses, churches, and public buildings, build roads, and crush ore and amalgamate silver. Reliance on the indigenous and enslaved African labour force as well as

[7] Or perhaps, as Millones Figueroa (2010) playfully speculates, the artists simply wanted Christ's final meal to be the most delicious.

migration from rural areas meant that cities would become multi-ethnic redoubts. Early seventeenth-century census data from Trujillo and Lima show substantial non-European populations, and as historian Karen Graubart (2007, 15–16) goes on to note, all these groups shared residential and commercial spaces. Among an array of occupations held by indigenous urbanites, domestic service was most common for girls and women, and this work sometimes shaded into selling groceries and preparing foods in the marketplace in town plazas. Women cooks and dealers predominated at the lower end of the food trade, some of whom rose to operate profitable *chicha* businesses (70–73).

And what was on offer in the plazas? Garofalo (2001) describes the lively social scenes in early seventeenth-century Cuzco and Lima where vendors proffered fish, bread, fruits, eggs, nuts, dried peppers, medicinal herbs, and much more. These public spaces gathered people of all backgrounds with the elite looking on curiously, while diners of the humbler classes enjoyed meals served up by indigenous and Afro-Peruvian cooks, who combined culinary elements of three continents: "thick stews (locros) of corn or llama meat seasoned with aji: trout and boiled or toasted beans (especially on Catholic fast days): cooked potatoes and jerked llama and mutton seasoned with Andean pepper…white or purple com puddings (mazamorras), sometimes sweetened with sugar cane syrup or milk" (73). Similarly, a great variety of both European and Andean foods appeared in the shops and stalls of Potosí (DeFrance 2003). The sheer range of items from diverse production zones intimates the magnitude of the supply chains supporting urban markets.

The long-distance traders and petty entrepreneurs responsible for this bounty illustrate indigenous intervention in colonial markets, which as Stern (1995) points out had occurred in many forms since their beginning, as native peoples endeavoured to use Settler structures and practices to protect community resources and expand their own repertoire of livelihood options. This ongoing market engagement also belies the dichotomy between subsistence or "natural" and market economies sometimes mapped onto indigenous and European spaces. The interpenetration of economies is both concretized and symbolized in foodways; most of the meals I've enjoyed in rural Highland hamlets and homes featured a hefty helping of rice below or next to mutton or alpaca or tuna or a fried egg.

Further, the mobilities of individuals across rural and urban settings, like the residential mosaic of multicultural seventeenth-century cities, mitigate or at least nuance other pervasive dualisms: the separate spheres of Spaniard/Indian, Coast/Highland, and wheat/maize. The point in respect to gastronomic hybridities is simple: through ongoing interactions, in beds, households, workshops, and markets, indigenous peoples, Africans, and Spaniards and the various combinations of their offspring had reason to share plazas and sometimes hearths and become familiar with each other's foods, especially as those very categories of "other" have shifted, mutated, and blurred and been renamed and redeployed over the decades and centuries.[8]

Concluding Thoughts

Ending the account of colonial dietary hybridity here would in part reprise the story told by authors of the gastronomic boom, of a multicultural feast that brings all citizens together at a table of reconciliation and prosperity. Although transcultural recipe exchange continues through the end of the colonial era (in 1821), and into the present century, it occurs within the structures of domination set in place with conquest. As the aforementioned critics and sceptics of contemporary gastropolitics have shown, that story of delicious conviviality omits many chapters and elides chasms of wealth and power among the characters. Even gastronomy can serve as another settler frontier and the appropriation of indigenous foodways as another dimension of colonial extraction and subordination of native heritage (Grey and Newman 2018).

I close with a caution about the legacies of conquest that today roil Peru. In his argument for the inclusion of Latin American countries as settler societies, Richard Gott (2007) highlights as characteristic a sustained elite fear of and contempt for the ethnic underclass, something manifest across the region in nineteenth-century efforts to whiten

[8] A discussion of Andean ethnicity in all its diachronic and spatial fluidity is far beyond the scope of this chapter. A small sampling of scholarly treatment includes Babb (2020); Cahill (1994); de la Cadena (2009); Thorp and Paredes (2010); and Takenaka (2004).

respective national populations while securing labour by encouraging European immigration.[9] Gott explains that his ideas about Latin America found their genesis in his observations of the virulently racist responses to the Cuban Revolution, Allende's election in Chile, and Hugo Chavez's regime in Venezuela, and later in reactions to the Pink Tide of the aughts. In these cases, radical power shifts implied an advancement for people of colour, one that prompted "white protest and panic" (271). Just these kinds of ideological currents have dramatically emerged in recent months as the Peruvian presidential election pitted Keiko Fujimori against Pedro Castillo. While perhaps neither is a classic avatar of historic divisions, Castillo is identified as the candidate of the Left and of indigenous aspirations, while Fujimori, daughter of the imprisoned former president, has had overwhelming support among the upper classes. As of this writing in the late Spring of 2021, Castillo, the winner, has been subject to a barrage of conservative hyperbole and overtly racist attacks, as well as with the opposition's efforts to overturn the results (Collyns 2021). This ugly contest seems far removed from the optimistic, even redemptive discourses of Peruvian gastropolitics; nonetheless, it alerts us to the festering tensions too easily concealed within. Food and foodways carry stories about and for their human progenitors, and the episodes related here can perhaps shed light on the unsettling present.

[9] This effort largely failed in Peru as Europeans rejected the incentives offered by the Peruvian government, leading it, grudgingly, to permit Japanese immigration (Takenaka 2004). Even earlier, a much larger number of Chinese labourers had been brought to Peru to replace enslaved Africans mining guano and in agriculture (Palma and Ragas 2019). While the social paths of these two groups differ, their foodways have been embraced within the gastronomic boom. The Japanese migrants who survived the harsh conditions of their indentured servitude on coastal plantations largely headed to Lima. Through mutual aid and ethnic solidarity, they succeeded in opening small businesses, notably restaurants and bakeries. Their mid-century success and visibility made them targets for intensified racism and violence (Takenaka 2004). For contemporary Japanese Peruvians, the celebration of Nikkei foods represents national acceptance (Takenaka 2019).

Bibliography

Aldenderfer, Mark. 2001. *Andean Pastoral Origins and Evolution: The Role of Ethnoarchaeology*. In *Ethnoarchaeology of Andean South America: Contributions to Archaeological Method and Theory*, ed. Lawrence Kuznar. New York: Berghahn Books.

Álvarez, Isabel. 2007. Panorama de la concina Peruana el siglo XX y perspectivas para el siglo XXI. In *Seminario Histórico de la Cocina Peruana*, ed. M. Villavicencio, 243–356. Lima: Universidad de San Martín de Porres.

Babb, Florence E. 2020. 'The real indigenous are higher up': Locating Race and Gender in Andean Peru. *Latin American and Caribbean Ethnic Studies* 1-22. https://doi.org/10.1080/17442222.2020.1809080.

Cahill, David. 1994. Colour by Numbers: Racial and Ethnic Categories in the Viceroyalty of Peru, 1532–1824. *Journal of Lain American Studies* 26 (2): 325–346.

Collyns, Dan. 2021. Peru: Fujimori Cries Electoral Fraud – And Unleashes Torrent of Racism. *The Guardian, June* 20: 2021. https://www.theguardian.com/world/2021/jun/20/peru-elite-election-pedro-castillo-keiko-fujimori.

Contreras, Catherine. 2011. ¿Cómo Lima se convertirá en capital gastronómica de América en el 2021? *El Comercio*, September 10. Accessed October 4, 2011. http://elcomercio.pe/gastronomia/1288535/noticia-como-lima-se-convertira-capital-gastronomica-america-2021.

Cook, David Noble. 1998. *Born to Die: Disease and New World Conquest 1492–1650*. New York: Cambridge University Press.

Crosby, Alfred W. 2003. *The Columbian Exchange: Biological and Cultural Consequences of 1492*. Greenwood Press 1972: Praeger Publishers.

D'Altroy, Terence N. 2015 Funding the Inca Empire. In The Inka Empire: A Multidisciplinary Approach. Izumi Shimada, ed. 97–118. Austin: University of Texas Press.

Davis, Keith. 1984. *Landowners in Colonial Peru*. Austin, TX: University of Texas Press.

De la Cadena, Marisol. 2009. *Indigenous Mestizos: The Politics of Race and Culture in Cusco, Peru, 1919–1991*. Durham, NC: Duke University Press.

deFrance, Susan, and D. 2021. Guinea Pigs in the Spanish Colonial Andes: Culinary and Ritual Transformations. *International Journal of Historical Archaeology* 25 (1): 116–143. https://doi.org/10.1007/s10761-020-00548-6.

deFrance, Susan D. 2003. Diet and provisioning in the high Andes: a Spanish colonial settlement on the outskirts of Potosí, Bolivia. International Journal of Historical Archaeology 7: 99–125.

Doughty, P. 2001. Ending Serfdom in Peru: The Struggle for Land and Freedom in Vicos. In *Contemporary Cultures and Societies of Latin America: A Reader in Social Anthropology of Middle and South America*, 3rd Edition. ed. Dwight Heath, 222–243. Waveland Press: Prospect Heights, IL.

Earle, Rebecca. 2012. The Body of the Conquistador. NY: Cambridge University Press.

Figueroa, Luis Millones. 2010. The Staff of Life: Wheat and 'Indian Bread' in the New World. *Colonial Latin American Review* 19 (2): 301–322. https://doi.org/10.1080/10609164.2010.493688.

Flores Ochoa, Jorge. 1988. Cambios en la Puna. In *Llamichos y Paqocheros: Pastores de Llamas y Alpacas*, ed. Jorge Flores Ochoa, 273–293. Cuzco: Centro de Estudios Andinos.

García, María Elena. 2013. The Taste of Conquest: Colonialism, Cosmopolitics, and the Dark Side of Peru;s Gastronomic Boom. *The Journal of Latin American and Caribbean Anthropology* 18 (3): 505–24.

———. 2021. Gastropolitics and the Specter of Race: Stories of Capital, Culture, and Coloniality. Berkeley: University of California Press.

Garofalo, Leo. 2001. The Ethno-Economy of Food, Drink, and Stimulants: The Making of Race in Colonial Lima and Cuzco. PhD, History, University of Wisconsin.

Gott, Richard. 2007. Latin America as a White Settler Society. *Bulletin of Latin American Research* 26 (2): 269–289. https://doi.org/10.1111/j.1470-9856.2007.00224.x.

Graubart, Karen. 2007. *With our Labor and our Sweat: Indigenous Women and the Formation of Colonial Society in Peru, 1550–1700*. Stanford, CA: Stanford University Press.

Grey, Sam, and Lenore Newman. 2018. Beyond Culinary Colonialism: Indigenous Food Sovereignty, Liberal Multiculturalism, and the Control of Gastronomic Capital. *Agriculture and Human Values* 35: 717–730.

IPSOS Public Affairs. 2020. *I Encuesta Nacional sobre Diversidad Cultural y Discriminación*. Lima.

Janer, Zilkia. 2007. (In)Edible Nature. *Cultural Studies* 21 (2/3): 385–405. https://doi.org/10.1080/09502380601162597.

Kubler, George. 1946. The Quechua in the Colonial World. In Handbook of South American Indians: The Andean Civilizations, Vol 2. Julian Steward, ed. 331–410. Washington, DC: Smithsonian Institution.

Khalip, Andrei. 2007. Peru's Capital a Destination for Foodies; Tours of Lima Showcase Newly Popular Cuisine. *Washington Post* 25: A11.

Klaren, Peter. 2005. The Sugar Industry in Peru. *Revista de Indias* 65 (233): 33–48.

Kollenda, Heidrun. 2019. From Farm to Table: Productive Alliances as a Pathway to Inclusive Development in Peru. *Anthropology of Food* 14. https://doi.org/10.4000/aof.9992.

Moore, Katherine. 2016. Early Domesticated Camelids in the Andes. In The Archaeology of Andean Pastoralism. Capriles, José M. and Nicholas Tripcevich, ed. 17–38. Albuquerque: University of New Mexico Press.

Nair, Stella and Jean-Pierre Protzen. 2015. The Inka Built Environment. In The Inka Empire: A Multidisciplinary Approach. Izumi Shimada, ed. 215–31. Austin: University of Texas Press.

Krogel, Alison. 2011. *Food, Power, and Resistance in the Andes: Exploring Quechua Verbal and Visual Narratives*. Lanham: Lexington Books.

Taj, Mitra and Anatoly Kurmanaev. 2020. Virus Exposes Weak Links in Peru's Success Story. *New York Times*, June 12. https://www.nytimes.com/2020/06/12/world/americas/coronavirus-peru-inequality-corruption.html.

Livi-Bacci, Massimo. 2006. The Depopulation of Hispanic America After the Conquest. *Population and Development Review* 32 (2): 199–232.

López-Canales, Jorge. 2019. Peru on a Plate: Coloniality and Modernity in Peru's High-End Cuisine. *Anthropology of Food* 14. https://doi.org/10.4000/aof.10138.

Marcus, Joyce, Jeffrey D. Sommer, and Christopher P. Glew. 1999. Fish and Mammals in the Economy of an Ancient Peruvian Kingdom. *Proceedings of the National Academy of Sciences* 96 (11): 6564–6570. https://doi.org/10.1073/pnas.96.11.6564.

Markowitz, Lisa. 2012. Highland Haute Cuisine: The Transformation of Alpaca Meat. *Re-Imagining Marginalized Foods. Elizabeth Finnis*, ed: 34–48.

———. 2018. Making and Unmaking the Andean Food Pyramid: Agronomy, Animal Science, and Ideology. In *The Andean World*, ed. Linda Seligmann and Fine-Dare, Katherine, 205–218. New York: Routledge.

Matta, Raúl. 2016. Food Incursions into Global Heritage: Peruvian Cuisine's Slippery Road to UNESCO. *Social Anthropology* 24 (3): 338–352. https://doi.org/10.1111/1469-8676.12300.

McDonell, Emma. 2019. Creating the Culinary Frontier: A Critical Examination of Peruvian Chefs' Narratives of Lost/Discovered Foods. *Anthropology of Food* 14. https://doi.org/10.4000/aof.10183.

McLaughlin, Katy. 2011. The Next Big Thing: Peruvian Food. *The Wall Street Journal*, September 10. http://online.wsj.com/article/SB10001424053111904199404576540970634332968.html.

Morner, Magnus. 1985. *The Andean Past: Land, Societies, and Conflicts*. New York: Columbia University Press.

Murra, John V. 1980. *The Economic Organization of the Inka State. Research in Economic Anthropology: Supplement 1*. Greenwich, CT: JAI Press.

National Research Council. 1989. *Lost Crops of the Incas: Little Known Plants of the Andes with Promise for Worldwide Cultivation*. Washington, DC: National Academy Press.

Nava, Armando. 2015. Inequality in Peru: Reality and Risks. In *Oxfam Working Paper*. https://cng-cdn.oxfam.org/peru.oxfam.org/s3fs-public/file_attachments/Inequality%20in%20Peru.%20Reality%20and%20Risks_4.pdf.

Newson, Linda A. 1993. The Demographic Collapse of Native Peoples of the Americas, 1492–1650. *Proceedings of the British Academy* 81: 247–288.

Ortiz, Marienella. 2010. El boom de nuestra gastronómica debe beneficiar al agro. *El Comercio*, March 29. Accessed April 1, 2010. http://elcomercio.pe/noticia/453611/boom-nuestra-gastronomia-arrastrar-al-agro-hacia-crecimiento.

Palma, Patricia, and José Ragas. 2019. Feeding Prejudices: Chinese Fondas and the Culinary Making of National Identity in Peru. In *American Chinese Restaurants: Society, Culture and Consumption*, ed. Jenny Banh and Haiming Liu, 44–61. London: Routledge.

Passidomo, Catarina. 2017. Our Culinary Heritage: Obscuring Inequality by Celebrating Diversity in Peru and the U.S. South. *Humanity & Society* 41 (4): 427–445.

Patel, Raj. 2021. Food Injustice Has Deep Roots: Let's Start with America's Apple Pie. *The Guardian*, May 1, 2021. https://www.theguardian.com/environment/2021/may/01/food-injustice-has-deep-roots-lets-start-with-americas-apple-pie.

Peruvian Ministry of Agriculture. 2008. *Native Potatoes of Peru: Catalogue of Varieties and Culinary Uses*. Lima: Ministry of Agriculture.

Rostworowski, Maria. 2007. La Cocina Prehispánica. In *Seminario Histórico de la Cocina Peruana*, ed. M. Villavicencio, 120–129. Lima: Universidad de San Martín de Porres.

Rovzar, Chris. 2016. *Food Tourism Could Bring $7.5 Billion to Peru in 2016* (Original edition). Bloomberg.

Speed, Shannon. 2017. Structures of Settler Capitalism in Abya Yala. *American Quarterly* 69 (4): 783–790.

Stein, Stanley, and Barbara Stein. 1970. *The Colonial Heritage of Latin America. Essays on Economic Dependence in Perspective*. New York: Oxford University Press.

Stern, Steve. 1982. *Peru's Indian Peoples and the Challenge of Spanish Conquest: Huamanga to 1640*. Madison, WI: University of Wisconsin Press.

———. 1995. The Variety and Ambiguity of Native Andean Intervention in European Colonial Markets. In *Ethnicity, Markets, and Migration in the Andes: At the Crossroads of History and Anthropology*, ed. Brooke Larson and Olivia Harris, 73–100. Durham, NC: Duke University.

Super, John. 1988. *Food, Conquest, and Colonization in Sixteenth-Century Spanish America*. Albuquerque: University of New Mexico Press.

Takenaka, Ayumi. 2004. The Japanese in Peru: History of Immigration, Settlement, and Racialization. *Latin American Perspectives* 31 (3): 77–98. https://doi.org/10.1177/0094582x04264745.

———. 2019. "Nikkei Food" for Whom? Gastro-Politics and Culinary Representation in Peru. *Anthropology of Food* 14. https://doi.org/10.4000/aof.10065.

Taylor, Lucy. 2020. Four Foundations of Settler Colonial Theory: Four Insights from Argentina. *Settler Colonial Studies*. https://doi.org/10.1080/2201473X.2020.1845939.

Thorp, Rosemary, and Maritza Paredes. 2010. *Ethnicity and the Persistence of Inequality: The Case of Peru*. London: Palgrave.

Wernke, Steven. 2013. *Negotiated Settlements: Andean Communities and Landscapes under Inka and Spanish Colonialism*. Gainesville, FL: University Press of Florida.

Wolfe, Patrick. 2006. Settler Colonialism and the Elimination of the Native. *Journal of Genocide Research* 8 (4): 387–409.

3

What Belongs in the "Federal Diet"? Depictions of a National Cuisine in the Early American Republic

Peter Mabli

Introduction

Defining a national cuisine in the early American Republic was in many respects a fool's errand. Indeed, the development of a cohesive, singular, and unified cuisine remains amorphous and problematic in the United States today. But these difficulties did not stop Americans from attempting to codify their eating habits and eschewing such a concept's importance in cementing a burgeoning national identity separate from their European colonial predecessors. The production, dissemination, and consumption of certain foodstuffs were essential components in characterizing the spirit of a settler-colonial (and later national) consciousness.

But the attempted development of this national American cuisine was often exclusionary and seemingly arbitrary; it was a process that ignored certain peoples and their culinary traditions while embracing others. While identity formation through food choices in the late colonial period

P. Mabli (✉)
American Social History Project, The Graduate Center, CUNY,
New York, NY, USA

R. Ranta et al. (eds.), *'Going Native?'*, Food and Identity in a Globalising World,
https://doi.org/10.1007/978-3-030-96268-5_3

and early national period allowed Americans of European ancestry to recognize food as a powerful cultural and political tool, there was a need, as critical theorist Homi Bhabha writes regarding more recent postcolonial societies, to manifest a "desire for a reformed, recognizable Other, as a subject of a difference that is almost the same, but not quite" (Bhabha 2004, 112). In this new identity, indigenous peoples[1]—whose foodstuffs remained essential components of a new "American" cuisine—were often rejected from the discourse and ignored in any of its definitions.

Beginning early in the colonial period, European settlers detailed the impressive natural resources of the North American continent. Over generations, these resources were slowly and consciously removed from their indigenous origins to disassociate their surroundings from "savage" lifestyles. In the decades preceding the American Revolution, colonists developed a complex food system that helped define their identity, an identity consciously—and violently—unlinked from the climate, culture, and people of the continent. As Walter Hixon writes, "Indian removal and indiscriminate warfare evolved in tandem with the formation and achievement of American racial and nationalist aspirations… In the face of this unifying discourse, colonial ambivalences degenerated into campaigns of ethnic cleansing" (Hixon 2013, 55). American cuisine of the era was, therefore, created more through acculturation and destruction than unification and originality: Americans absorbed dishes that originated in other cultures and fit them into their own national history. They rejected certain European, African,[2] and Native American culinary characteristics and embraced others to create a distinct and hybridized culinary culture.

---

[1] When discussing ingredients and foodstuffs found in North America, I use the descriptor "native" to differentiate items from their European counterparts. The North American people with ancestral claims in the continent prior to European contact, however, are described as "indigenous" peoples or the descriptor "Native Americans." While there are other equally valid terms used to describe this population, I use these terms because of their frequent use in academic writing in the United States.

[2] While this chapter will focus on the American settler-colonial responses to and dismissal of certain Native American foodways, there are many other instances of similar actions perpetrated on other peoples and their cuisines. The white American response to African and African American foodways is a particularly poignant example of excluding a people while simultaneously embracing certain components of their cuisine.

"A Federal Diet" and the Prospects of a Unified Nation

The attempts to formulate a codified national diet in the pages of American newspapers in the late 1780s are an example of this complex, exclusionary construct. An article entitled "A Federal Diet" in the October 8, 1788, edition of *The Philadelphia Gazette* addressed the inherent complexities of culinary identity formation when it provocatively declared: "Solomon, speaking of mankind, says, 'madness is in their *hearts*, had he lived in the United States...he would certainly have added, that madness is in their *stomachs*.'" The supposed "madness" was caused by the prominence of imported foodstuffs in the American people's dietary habits, a practice by which Americans obtained "their *breakfast* from China, their *drinks* from the West India islands, and their *supper* (when it consists of *cheese*) from Great Britain." According to the author, reliance on the importation of daily necessities had so poorly affected the culture that this "American malady" needed to be immediately cured with a radical transformation and reinvention of culinary habits—a new system of food production and consumption throughout the country that he named the "federal diet" (Anonymous, "A Federal Diet").

The article was reprinted in numerous newspapers throughout the country after its initial publication on October 8. Without stringent copyright laws, it was a common practice for newspaper editors throughout the United States to reprint articles from other sources. Generally, if the original paper was referenced or if the article came from exchange papers (newspapers with a countrywide subscription audience that published general correspondences), it was an accepted practice to not credit the original printer (Pasley 2002, 77). It is significant therefore that the "federal diet" article, which was not from an exchange paper, was reprinted in at least eight other newspapers in seven states from October to November 1788.[3] The frequency of reprints, the scope and range of the

[3] "A Federal Diet" was reprinted in *The Pennsylvania Packet and Daily Advertiser* on October 9, 1788; *The Connecticut Journal* on October 15, 1788; *The New-Jersey Journal and Political Intelligence* on October 15, 1788; *The Massachusetts Centinel* on October 18, 1788; *The American Mercury* (Philadelphia, PA) on October 20, 1788; *The United States Chronicle; Political, Commercial and Historical* (Providence, RI) on October 30, 1788; *The Freeman's Oracle Or New-Hampshire Advertiser*

publications, and the relative speed with which the letter was reprinted suggests an acknowledgement of the importance of its content by certain editors—an article of greater importance than the more conventional opinion pieces advocating for a stronger federal system of government.

What was it that made the "federal diet" editorial so influential? Certainly the timing of its publication played a role in its numerous reprints. By the summer of 1788, all but North Carolina and Rhode Island had ratified the U.S. Constitution, setting the stage for Congress to meet and complete the process the following March 1789. Discussions on the scope and influence of federalism crowded the newspapers and debates of the era as Americans grappled with what it would truly mean to be a country united under a more powerful federal system of government. This unifying federal concept was indicative of a dramatic cultural shift in American society that prioritized the formation of institutions based on republican ideals in order to enlighten the American citizenry and encourage a sense of civic unity throughout the country.

While arguments over federal reach, citizenship, and the national economy were mainstays of American discourse in the years surrounding the ratification process, the influence such a federalist system would have on the quotidian habits of the American people was rarely reviewed. The concept of a "federal diet," therefore, was a novel idea that spoke to the overwhelming reach that a new, postcolonial identity could have on not only the political and economic realms of American culture, but also the daily lives of its citizens. The article was a deliberate and codified attempt to define the nascent nation using food culture as a unifying social system. What Americans ate, therefore, could unite them as strongly as their political and economic institutions.

More than simply a call to action, the "federal diet" letter continued with a specific plan for the implementation of an independent postcolonial culinary system. The letter detailed the preparation methods and types of foods that Americans should eat, divided into sections for breakfast, dinner, and supper. For breakfast, "milk may be used in a variety of

on November 1, 1788; and *The Vermont Gazette* on November 24, 1788. Of these papers, only four papers give proper credit to *The Pennsylvania Gazette. The New-Jersey and Political Intelligence* went as far as to omit the greeting "Mess'rs Hall & Sellers" from the original article—the printers of *The Pennsylvania Gazette*—and added a greeting to their own printer.

ways...either with bread, or rice, or Indian mush, or apple pye, or with the ripe fruits of the season." That milk—a drink used primarily to "fatten hogs" in the early national period and not as often for human consumption—should be a primary ingredient in the American breakfast spoke to the author's desire to defy certain cultural stigma and promote resourcefulness and independence in American society. Furthermore, the choice between bread, rice, or "Indian mush" as a grain reveals the author's awareness of different food staples prevalent throughout the country (short grain white rice in particular being a staple of southern states and low-country diets) and speaks to his attempt to unite Americans while remaining cognizant of their regional preferences. And choosing to eat "ripe fruits of the season" further allowed for regional variations as well as championed fresh, local produce as opposed to imported dried or preserved products.

In contrast to the other sections, the author left open the options for dinner. For this meal, "all the meats and fishes of our country, with bread made of wheat, rye or Indian corn," were acceptable, "the more vegetables that are eaten with meat the better." No other specifics were offered. Instead, the author devoted the remaining space to a discussion on the cost savings of rye bread versus wheat and the health benefits of vegetable broth, particularly for younger children. Supper, he insisted, should consist of simply "the same as breakfast." The generality of the final section either suggests the author's inability to find a single comprehensive foodstuff agreeable to all Americans for supper or was an attempt to promote the variety and abundance of American foods by allowing citizens to decide on their own most appealing menu options. Regardless, it is regrettable that the letter ends with such little detail and without a compelling conclusion (Anonymous, "A Federal Diet").

What the "federal diet" letter did accomplish, however, was the creation of a plan to bridge the gap between a dependent colonial culture and a postcolonial republican society by detailing a set of culinary guidelines for all Americans to follow. By advocating American-made products with priority given to local produce, the author attempted to compel the country towards self-reliance and conviviality through the consumption of a unified singular diet.

In the weeks following its initial publication, *The Pennsylvania Gazette* and other newspapers published follow-up articles that complimented and amended the call for a "federal diet." One of the more detailed articles contained instructions to convert the sap of maple trees into a variety of useful foods and drinks (Agricola, "To Make Maple Sugar"). Echoing the "federal diet's" concern with an over-dependence on imported goods, the author, named "Agricola"—the Latin word for "farmer"—believed the "immense sums of money sent every year to the West-Indies for sugar" could be used instead to cultivate "a species of the American maple containing genuine sugar, and if properly prepared, would in every respect equal in all its qualities the sugar obtained from the cane of the West-Indies." The author saw a great deal of potential from the effective cultivation of certain American foods, particularly in relation to their international counterparts.[4] What was lacking was not the quantity or quality of the natural resource (milk for the cheese and trees for the maple syrup) but the will of American farmers and merchants to adopt new procedures to effectively prepare these foodstuffs and implement improved methods of production.

The article continued with a list of recipes made using the sap of trees, each "earnestly recommended to those citizens of the United States who live in the neighbourhood of the sugar maple" (Rush 1792, 10). The items ranged from a simple maple sugar made from boiling and refining maple sap to maple molasses and even maple beer, wine, and vinegar. An impressively detailed description of each step accompanied the items, allowing readers to easily recreate and attempt each recipe with their local supply of maple trees. Thus, the increased awareness of the confluence of republican federalism and postcolonial culture during the late 1780s—particularly in regards to food consumption—is evident not only in the "federal diet" and its accompanying editors but also in other published articles of the period.

[4] Dr Benjamin Rush would echo this sentiment in a widely published treatise on maple sugar trees. "The *quality* of this sugar is necessarily better than that which is made in the West-Indies," he wrote, "whoever considers that the gift of the sugar maple trees is from a benevolent Providence, that we have many millions of acres in our country covered with them, that the tree is improved by repeated tappings, and that the sugar is obtained by the frugal labour of a farmer's family…will not hesitate in believing that the maple sugar may be manufactured much cheaper, and sold at a *less price* than that which is made in the West-Indies" (Rush 1792, 9–10).

The Omission of Yaupon from "A Federal Diet"

One of the more problematic foodstuffs mentioned in the "federal diet" system were caffeinated drinks, for which the author advocated complete abstinence. Tea and coffee, however, were mainstays of American tables for generations, and the national wholesale abandonment of these drinks would be extremely impractical. Yet in clear terms, the "federal diet" advocated for their removal. The drinks are only mentioned under the "breakfast" section of the essay, oddly referenced in passing when discussing salted meats: "Salt meat of all kinds, and salt fish (which has been made a part of the American breakfast in order to make up for the deficiency of *tea* and *coffee* articles of diet) should never be tasted at breakfast." Unfortunately, how salted fish supposedly became a substitute for coffee and tea is not discussed any further.

Nevertheless, the strict abandonment of tea and coffee is an obvious conclusion of the proposed system, as according to the rules laid out by the author, imported items are to be avoided at all costs. What is of note, however, is that the author provided no suitable native-based alternative to the consumption of coffee and tea.[5] With the use of the sap of North American maple trees, copious research and pages of recipes were presented to showcase a viable alternative to imported sugar. Americans following the "federal diet," however, were left with no substitute for a caffeinated breakfast beverage, when in fact there did exist a native alternative available.

The solution could lay in yaupon (also known as cassina), a species of holly native to the southeastern United States. The leaves of the plant contain high levels of caffeine, and it is the only North American holly species to contain high enough levels of the drug to cause a noticeable chemical reaction in humans. Yaupon leaves had been dried, boiled, and drunk by the Timucuan peoples of modern-day Florida for generations

[5] There were a number of concoctions made to substitute imported tea during the 1760s and 1770s (often collectively called "liberty teas") made from ground flax and other local flowers and dried herbs. None were caffeinated, however, making them rather ineffective as tea substitutes. For a detailed description of these "liberty teas" as well as the impact of boycotts on domestic life in colonial and revolutionary America, see: Siegel, Nancy. 2008. "Cooking Up American Politics," *Gastronomica* 8, no. 3: 53–61.

prior to the arrival of European colonists in North America, and its use as a tea was well known by European colonists as early as the mid-sixteenth century (Pearson 1972).

During explorer Jean Ribault's 1564 expedition to the southeastern corner of North America, crewmember and artist Jacques le Moyne was the first European to depict numerous Native American rituals. One of his engravings contained an event described as a Timucua "council of state" in which "the chief sits at the place of honor surrounded by his advisors," and "a bitter black tea called casina [*sic*] is prepared by the women" (Moyne 1564) (Fig. 3.1).

In the image, a woman is seen dropping leaves into large boiling cauldrons, while another ladles the mixture and pours it through a sieve into another pot. Large conch shells are passed around containing the tea to the council members, as French Huguenots (clad in armour and brandishing muskets and pikes) look on. These engravings were reproduced

Fig. 3.1 Moyne, Jacques le. *Plate XXIX: A Council of State*. 1564. Engraving. Special Collections Department. University of South Florida Archives

and compiled with maps and other depictions of the New World by master engraver Theodorus de Bry, who widely disseminated them throughout Europe in his publication, *America*.

Yaupon tea continued to be consumed in portions of the southern United States well into the nineteenth century. An 1842 botany guide described the yaupon fruit as being "used for tea in Paraguay, and the famous *black drink* of the Creek Indians is prepared from the leaves of cassena; which are still used as a substitute for tea in some parts of the Southern States" (Gray 1842, 308). A later essay by chemist Henry M. Smith described yaupon as having a "very agreeable odor, perhaps faintly resembling that of raw tobacco, but having also a tea-like smell" and was "largely used in the South as a substitute for tea, coffee, and other stimulants" (Smith 1872, 216–217). Thus, yaupon was a well-known alternative to imported caffeinated drinks such as tea and coffee and had been widely consumed for centuries by Native American, European, and American populations throughout large sections of North America.

Why then, would the proponent of the "federal diet" present maple syrup as a viable substitute for imported sugar, but not present a drink such as yaupon as a viable substitute for imported tea and coffee? Regional variation may have been partly to blame, as yaupon was mostly consumed by peoples in the southern United States. However, as previously noted, recommendations in the "federal diet" did consist of other southern regional staples such as cultivated short grain rice. It is therefore probable that the "federal diet" could incorporate other regional items like yaupon, if the food item were considered a beneficial addition. Further, yaupon is a light and dried foodstuff that (like tea and coffee) could have been transported over long distances without the fear of spoil, allowing for countrywide consumption and effectively negating its regional nature as an obstacle.

It is more likely a certain cultural stigma associated with the drink that barred it from being adopted on a national level. Yaupon tea was erroneously labelled an emetic by early European settlers: early depictions of the drink associated it with warrior endurance rituals that often ended in the participants vomiting. The first example of such a depiction was Jacques le Moyne's image discussed previously, which showed two men sitting around a fire circle vomiting while others nearby drank the tea (Moyne 1564).[6] It was a false correlation, however: the tea does not make a person vomit, and eyewitness testimony—either through mistranslation or through misconstrued cultural cues—confused the participants' reactions to the physical exhaustion of the ritual with imbibing yaupon. Nevertheless, the association of yaupon with vomiting became so ingrained in Western culture that well into the twentieth century, the scientific term for the yaupon holly remained *Ilex vomitoria* (Power and Chesnut 1919, 1308).

The example of yaupon highlights an important requisite for the appropriation of specific foodstuffs into the American dietary lexicon. While both maple trees and yaupon holly are native North American species with well-documented culinary uses, yaupon tea's misrepresentation and close association with Native American society prevented its assimilation into European settler—and later colonial and post-revolutionary American—culinary traditions. Although the conversion of maple sap into edible syrup was a practice first introduced to European settlers by indigenous peoples throughout the northern forests of North America, the production methods and byproducts developed were far enough removed from their indigenous origins that maple syrup no longer bore the stigma of a foodstuff solely consumed and produced by indigenous peoples (Mower 2007, 365). By the eighteenth century, European colonial maple sugaring methods had become a process distinct from their Native American counterparts. Copper cauldrons and piping replaced the wooden buckets used by indigenous peoples, and boiling to form the syrup reduction replaced the Native American technique of sap-freezing.[7] Consequently, the maple byproducts reflected a greater European culinary influence, such as the "maple beer" and "maple vinegar" recipes detailed in *The Pennsylvania Gazette*. Yaupon, however, remained associated with

[6] The descriptor of the plate stated, "Those who vomit up the tea are considered to be unfit for battle."

[7] The laborious process of sap-freezing consisted of draining the maple sap into wooden buckets, waiting for the sap to freeze, and scraping off the top layers of ice that would form. This process had to be repeated multiple times to completely remove the water from the sap and reduce the concoction into a maple syrup. This method was time-consuming as it forced participants to refreeze the sap over many nights and was dependent on seasonal cold temperatures that facilitated the freezing. The colonial preference for boiling was far more efficient and less time-consuming.

indigenous culture, and colonists' dismissiveness and negative attitudes towards indigenous people and society may have played a part in preventing the tea's assimilation into the federal American diet.

Separating Native Foods from Indigenous Peoples

European settlers and their colonial descendants' negative perceptions that Native American societies were supposedly "savage" remained prevalent during the early national period. The decision to incorporate certain indigenous foodstuffs into the "federal diet" hinged substantially on the ability of Americans to separate indigenous *foods* from their indigenous *cultures*. With an estimated 1800 distinct plant species used as food by indigenous peoples that were originally unknown to European colonists, the decision to incorporate one type of foodstuff over another was complex. Foods with similar counterparts in European cuisine (such as certain types of blackberries and cranberries) were at times easier to assimilate into colonial diets than completely foreign items. But familiarity was only one component; often colonists' disdain towards Native American society played a role in dietary decisions (Turner and Aderkas 2012, 311). A notable example of the European colonists' negative perceptions of Native American people and food culture is highlighted in Kariann Yokota's review of the 1807 children's book, *The People of All Nation; An Useful Toy for Girl or Boy*. The work compares a drawing of an indigenous Virginian holding tobacco leaves with a drawing of an Englishman eating a meat pie. Yokota writes, "Whereas New World peoples were associated with natural commodities, the 'civilized' Englishman in *The People of All Nations* is marked by his consumption rather than his production." Yokota concludes that this difference reflects the earlier Enlightenment concept that "a society's commercial production and habits of consumption were thought to reflect its 'character' and the level of civility it had attained" (Yokota 2011, 19–21). Indeed, while the "civilized" production and consumption of "savage" commodities remained an integral component of American identity formation well into the nineteenth century, it had its origins in early European colonists' prejudiced representations of Native American society and food culture.

The New World simultaneously mirrored the utopian land of milk and honey with a seemingly endless supply of raw materials and natural resources at their disposal, and a dangerous, ruthless wilderness that could ultimately destroy inhabitants' "civilized" identities. North America was, in many respects, the mythical medieval Land of Cockaigne made manifest.[8] As a result, American colonists soon learned to hybridize their eating habits by using native foodstuffs reminiscent of Old World counterparts or by cultivating when possible crops from Old World seeds. Although still dependent on a foreign environment to provide sustenance, colonists doggedly used their shared cultural identity to maintain cooking traditions. As historian William Cronon writes, "environment may initially shape the range of choices available to a people at a given moment, but culture reshapes environment in responding to those choices. The reshaped environment presents a new set of possibilities for cultural reproduction, thus setting up a new cycle of mutual determination" (Cronon 1983, 13). Through this acculturation process, colonial identities remained mostly intact, while the influence of Native American culture on colonial foodways was mitigated.

Even when praising the abundance of North American resources, for instance, some colonists had difficulty crediting Native Americans for their cultivars. David Pietersz de Vries, a Dutch settler in New Amsterdam, commented that in regard to the diet of the Native Americans, "the food supplies are various. The principal one is maize, which is their corn, and which is called by us Turkish wheat" (Vries 1909, 218). That de Vries and his fellow colonists would label corn as "Turkish" exemplifies the contempt with which they held Native Americans and their agriculture. For some European colonies, a grain so hearty, dependable, nutritious, and useful as corn could not be associated with "savage" peoples and therefore

[8] The Land of Cockaigne (an altered spelling of the Middle French term for "plenty") was a mythical realm where peasants had unlimited access to resources, including most notably unfettered access to food. While the land was normally described in a positive manner, it was feared by certain authors that the temptation of this lavish lifestyle would lead to idleness, and Cockaigne was ultimately considered harmful to its inhabitants. The concept first originated in the tenth-century poem, *The Land of Cockaygne*. See: Ginzburg, Carlo. 1992. *The Cheese and the Worms: The Cosmos of a Sixteenth-Century Miller*. Baltimore, MD: Johns Hopkins University Press, 84–86.

must have been a descendant of Old World seeds. Because food and communal identity remained well intertwined in colonial American mindsets, to remove corn from the continent was to remove Native Americans from corn.

Indeed, the belief that corn (arguably North America's most useful and profitable native crop) did not originate from North America remained a prevalent and much-debated biological theory well into the nineteenth century. In his essay on the history of corn, horticulturist Peter Arrell Browne wrote that as late as 1831, *Lieber's Encyclopedia Americana* argued, "the native country of the Indian corn remains still undetermined…it is a reflection upon, not only the learned, but the mass of the community, that their great *staple commodity*—the plant that demands and receives the patronage, the skill, and the industry of a large agricultural part of a great agricultural nation, should be so imperfectly known" (Browne 1837, 6). Browne debunked multiple theories that claimed corn originated in the Old World, including de Vries' assertion that corn was associated with Turkey. Browne's desire to rationally disprove the idea that corn was not native to the Americas spoke more to the negative colonial perceptions of Native American agriculture and society than to any scientific evidence needed to validate the contrary.[9]

Beyond discrediting indigenous populations' ownership claims to staple foods, European colonists also waged campaigns to replace indigenous crops with their own familiar foodstuffs. With varying levels of success, colonists grew crops that they were comfortable consuming and that would supposedly help maintain their collective health and identity. David Pietersz de Vries further contended that although there were some uses for "Turkish wheat," colonists preferred to harvest their own native foodstuffs:

> Our Netherlanders raise good wheat, rye, barley, oats, and peas, and can brew as good beer here as in our Fatherland, for good hops grow in the

---

[9] It is of note that Browne uses the term "Indian corn" to describe the foodstuff. By the nineteenth century, the term had won out over "Turkish grain" or any other geographically erroneous monikers, and "Indian corn" was the most widely used term. This change was an example of the slow evolutionary shift towards cultural acceptance of the crop's Native American origin in the ensuing decades.

> woods; and they who make it their business can produce enough of those things, as everything can be grown which grows in Holland, England, or France. It is a pleasant and charming country, if only it were well peopled by our nation. (Vries 1909, 219)

De Vries' pride for his fellow countryman is shown clearly in this passage. But more importantly, his insistence that foodstuffs grown in North America rivalled those of his homeland and other European nations signified a growing rift in colonial society: European colonists remained intellectually attached to their country of origin but increasingly began to make comparisons between their new home and their old, often ultimately favouring the new.

This sentiment was not isolated to the Dutch in New Amsterdam. Colonists in New England as well were grappled with similar effects of imposing European agricultural methods in North America. As historian Sarah F. McMahon argues, "Seventeenth-century colonists brought English plants, grains, and livestock to the New World, hoping to recreate their familiar fare in an unfamiliar country. Necessary alterations of the diet … reflected accommodations of old tastes and preferences to the agricultural realties of the new land" (McMahon 1985, 44). Indeed, throughout North America, European colonists formed a new culinary identity by begrudgingly adopting certain indigenous foodstuffs, removing their implicit connections to the indigenous people consuming them and obstinately cultivating traditional crops and re-creating European cooking methods. Though European by birth or by heritage, these descents of settler-colonists slowly saw the American continent as equal to (if not exceeding) Europe in resources and opportunities for future development. Abundant and bountiful crops and the high-quality food products produced from them were examples of the possibilities unique to America, but mostly in regard to those products acquired from the Old World and supplanted in the new. Embroiled within this newfound colonial identity formation, however, was a deliberate silencing of the indigenous foodstuffs and foodways—and by extension, the indigenous peoples—integral to the burgeoning cultural construct.

Conclusion

The years following American independence from Great Britain saw a dramatic and purposeful shift towards defining a postcolonial identity in more conclusive terms, leading to calls for systems such as the "federal diet," among other similar declarations. But declaring political independence and establishing the foundations for a national cultural identity in practical terms were not one and the same. Americans struggled to accurately define what it truly meant to be an American and began to come to terms with the difficulties inherent in restructuring a loosely connected and diverse citizenry into a singular entity.

By the end of the eighteenth century, American political and social structures had evolved into entities distinct from their European colonial counterparts. So too were American cultural systems slowly separating from their origins, including most notably the opinions and reactions towards American agriculture and diet. Early colonial reluctance to incorporate indigenous foodstuffs into their daily lives had gradually waned, and a new sense of pride and appreciation for the plentiful nature of the American continent (mostly void of descriptions of the indigenous populations themselves and their cooking styles) replaced previous misgivings. While such a national cuisine can instil a sense of pride in its adherents, the creation of a specific culinary identity had negative connotations for peoples deemed "outside" of the accepted populace. Indeed, ignoring or rejecting other cultures in order to espouse cultural dominance over the "other" was an integral component of national culinary identity formation.

By growing Old World crops on American soil and using traditional cooking methods, early European settlers helped differentiate themselves culturally from indigenous populations, even when native crops proved more practical to grow. The identities formed using food were often deductive; they found it easier and more effective to define what they were not (not British, not Native American, etc.) more so than what they actually had become. And in certain instances—such as the seemingly arbitrary choice to include maple syrup yet exclude yaupon tea in what may constitute a "federal diet"—what defined a singular American cuisine was affected by negative perceptions of certain cultures and peoples. Such ideas and methodologies lay the groundwork for the attempted codification and development of a distinct national cuisine, a concept that was predicated on both the rejection of colonial power structures and the removal of indigenous peoples from the narrative.

Bibliography

Bhabha, Homi K. 2004. *The Location of Culture*. 2nd ed. New York, NY: Routledge.

Browne, Peter Arnell. 1837. *An Essay on Indian Corn*. Philadelphia, PA: Printed by J. Thompson.

Cronon, William. 1983. *Changes in the Land: Indians, Colonists, and the Ecology of New England*. New York, NY: Hill and Wang.

Ginzburg, Carlo. 1992. *The Cheese and the Worms: The Cosmos of a Sixteenth-Century Miller*. Baltimore, MD: Johns Hopkins University Press.

Gray, Asa. 1842. *The Botanical Text-Book*. New York, NY: Wiley and Putnam Press.

Hixon, Walter L. 2013. *American Settler Colonialism*. New York, NY: Palgrave Macmillan.

McMahon, Sarah F. 1985. A Comfortable Subsistence': The Changing Composition of Diet in Rural New England, 1620–1840. *The William and Mary Quarterly* 42 (1): 26–51.

Mower, Robert M. 2007. Maple Syrup. In *The Oxford Companion to American Food and Drink*, ed. Andrew F. Smith. New York, NY: Oxford University Press.

Moyne, Jacques le. 1564. *Plate XXIX: A Council of State*. Engraving. Special Collections Department. University of South Florida Archives.

Pasley, Jeffrey L. 2002. *The Tyranny of Printers: Newspaper Politics in the Early American Republic*. Charlottesville, VA: University of Virginia Press.

Pearson, Fred Lamar. 1972. The Florencia Investigation of Spanish Timucua. *The Florida Historical Quarterly* 51 (2): 166–176.

Power, Frederick B., and Victor K. Chesnut. 1919. Ilex Vomitoria as a Native Source of Caffeine. *Journal of the American Chemical Society* 41 (8): 1307–1312.

Rush, Benjamin. 1792. *An Account of the Sugar-Maple Tree, of the United States, and of the Methods of Obtaining Sugar from It, Together with Observations Upon the Advantages Both Public and Private of This Sugar*. Philadelphia, PA: American Philosophical Society.

Siegel, Nancy. 2008. Cooking Up American Politics. *Gastronomica* 8 (3): 53–61.

Smith, Henry M. 1872. Yaupon. *The American Journal of Pharmacy* 4: 53.

Turner, Nancy J., and Patrick von Aderkas. 2012. Sustained by First Nations: European Newcomers' Use of Indigenous Plant Foods in Temperate North America. *Acta Societatis Botanicorum Poloniae* 81: no. 4.

de Vries, David Pietersz. 1909. Short Historical and Journal Notes of Various Voyages. In *Original Narratives of Early American History*, ed. John Franklin Jameson. New York, NY: Scribner.

Yokota, Kariann Akemi. 2011. *Unbecoming British: How Revolutionary America Became a Postcolonial Nation*. New York, NY: Oxford University Press.

4

The Taste of Colonialism? Changing Norms of Rice Production and Consumption in Modern Taiwan

Yujen Chen

Introduction

Among the existing global rice cultivars—Asian rice (Asian cultigen, Oryza sativa L.) and African rice (African cultigen, Oryza glaberrima Steudel)—Asian rice is cultivated more commonly around the world. Asian rice is further classified into Indica rice, Japonica rice, and Javanica rice, while Indica rice and Japonica rice each have their own glutinous rice subvariants. The three rice varieties are quite different in terms of property, suitable planting areas, and flavours. At present, Indica rice is most widely distributed. Specifically, it is mainly planted in Southeast Asia, Europe, and America. Japonica rice is primarily grown in Japan, South Korea, and northeast China, and Javanica rice is mainly planted in tropical mountains of Southeast Asia. However, it is interesting and unusual that the three rice varieties were once distributed and are still

Y. Chen (✉)
Department of Taiwan Culture, Languages and Literature, National Taiwan Normal University, Taipei City, Taiwan
e-mail: yjchen@ntnu.edu.tw

R. Ranta et al. (eds.), *'Going Native?'*, Food and Identity in a Globalising World,
https://doi.org/10.1007/978-3-030-96268-5_4

grown in Taiwan, an island without a considerable agricultural land area. The cultivation of the three rice varieties is closely related to Taiwan's immigration and colonization history. Rice is the staple food of the Taiwanese people, and its planting history in Taiwan also reflects Taiwan's colonization and immigration history over the past centuries.

Colonization and immigration have for the past 400 years been major influencing factors for food culture in Taiwan, as in other parts of the world subject to foreign conquest. However, the application of a settler-colonial framework to Taiwan is complicated by the fact that many Chinese settlers (mainland peoples of Han origin) arrived as migrants as well as agents of Dutch and later Qing conquest, while Taiwan's annexation by Japan was not premised on mass Japanese settlement of the island. Thus, while some traits of settler-colonial food cultures like land dispossession and appropriation or the imposition of metropolitan crops and tastes were present in Taiwan, others—such as the development of a distinctive and recognizable settler culinary repertoire—were not. As Hirano et al. (2018) have suggested, seen through a settler-colonial lens, Taiwan's modern history is that of 'dissonant and conflictive relationships among colonial authorities, settlers, and Indigenous peoples' which create an intersectional node of settler-colonial triangulation. In this sense, the Taiwanese case bears greater resemblance to the Mexican or Peruvian experiences discussed in this book, where colonialism radically transformed local foodways whilst simultaneously fostering creole or hybrid cuisines built on indigenous, colonial and immigrant contributions, rather than the South African or Australian experiences which delivered unique settler diets.

Before the seventeenth century, Taiwan was mainly inhabited by Austronesian aborigines, who planted staple crops, including millet and rice, as the main sources of food for the Formosan natives. In addition, taro and sweet potatoes were introduced to Taiwan before the seventeenth century, though the routes of entry were uncertain.[1] Among these crops,

[1] The Taiwanese aborigines did not leave any written documents before the seventeenth century. According to the earliest literature on Taiwan, authored by a Chinese officer, taro and sweet potato had been planted as supplements of staple food. See Chen, Di, *Dong Fan Ji [Writing Eastern Barbarians]*, (1603). Tsai and Yang (2004) suggested that sweet potatoes might be introduced to Taiwan from China or Southeast Asia.

millet had the most important function of ritual ceremonies in aboriginal societies. During the sixteenth and seventeenth centuries, Han people gradually migrated from coastal southeast China to establish agricultural settlements in Taiwan. From 1624 to 1662, Taiwan was among the commercial ports of the *Vereenigde Oostindische Compagnie* (VOC) (or the Dutch East India Company) in Asia. The VOC used Taiwan as its main stronghold to actively develop trade and agriculture, changing the farming modes and dietary habits of the aboriginal Taiwanese. The Dutch introduced many agricultural immigrants from the Chinese mainland, and many Han businesspeople served as intermediaries between the Dutch and aboriginal Taiwanese.

In 1662, the Dutch left Taiwan with the defeat by Zheng Chenggong (1624–1662), a general of the Ming Dynasty. Zheng's father was the most famous pirate in East Asia in the first half of the seventeenth century, and his mother was Japanese. Zheng Chenggong retreated to Taiwan because he was defeated by the troops of the Qing Dynasty (1644–1911). He, his sons, and grandchildren ruled Taiwan until they surrendered to the Qing Dynasty in 1683. During his reign in Taiwan, Zheng Chenggong implemented the *Tuntian* system[2] and developed agricultural settlements through encroachment or military conquest, resulting in a dramatic increase in the Han population. With the rapid increase of Han people, some aborigines moved to mountainous areas and others assimilated to Han Chinese culture and intermarried with Han Chinese. During the Qing Dynasty, the aborigines were officially classified as 'barbarians' (*Fan*), including 'wild' or 'raw' aborigines (*Sheng Fan*), and 'civilized' or 'cooked' aborigines (*Shou Fan*). The former refers to those who dwelt in mountainous regions and who did not pay tax, while the latter indicates those who were living in plains regions and who were better sinified, known as 'plains tribes' (*Pingpuzu*) (Li 1982).[3]

[2] The *Tuntian* system was a system of government-encouraged agriculture or agro-colonies that served to supply the colonists and military with agricultural products. In China, it can be traced to the Western Han dynasty (206 BCE–9 CE).

[3] According to statistics by the Interior Ministry, the population registering as aboriginals in 2020 was 571,816, accounting for 2.4% of the total population. See the Interior Ministry Website: https://www.ris.gov.tw/app/portal/346 (retrieved 9/30/2021).

The Qing Dynasty ruled Taiwan until 1895; it was defeated in the Sino-Japanese War, and Taiwan was ceded to Japan as a result. Taiwan, Japan's first overseas colony, was under Tokyo's rule until 1945. Subsequently, the Kuomintang government entered Taiwan along with more than one million immigrants, resulting in the remapping and transformation of food culture in Taiwan.

During the aforementioned historical changes, agriculture and diet have over hundreds of years been radically transformed. Specifically, the planting pattern and rice varieties serving as the staple food of the Taiwanese have undergone two major changes.

The first significant change was the large-scale planting of rice and the introduction of new rice varieties. For the aborigines, both millet and rice are essential foods (Tsai 2009: 16–17). In particular, millet is of great symbolic significance. Various legends about millet were circulated among many indigenous peoples, and millet played a vital role in sacrificial ceremonies.[4] Numerous Han immigrants arrived in Taiwan and were encouraged to plant rice through government policies. As a result, rice cultivation was significantly increased and surpassed that of millet. In addition to upland rice, paddy rice was gradually planted in Taiwan. Archives from the eighteenth century show that a few aborigines residing in the southern plains of Taiwan had learnt from the Han people how to build water management facilities for rice planting (Wang 1962 [1764]: 59, 82). However, aside from this area, few other aboriginal tribes had rice as their staple crop.

Plains Indigenous Peoples in southern Taiwan planted various rice varieties, including red glutinous rice and Javanica rice, while Indica rice could also be found in a few places (Tsai 2009). According to the records of Huang Shu-Ching (1680–1758), who was an officer of Qing government patrolling in Taiwan, the aboriginal Taiwanese began to plant rice

[4] Millet has been documented in the list of 'Ark of taste' of the Slow Food Foundation for Biodiversity. According to its description, Hafay—millet in Pangcah's language—is an ecotype of *Setaria italica* used to be one of the staple food for indigenous people in Taiwan, and its significance can be observed in varied traditional festivals of indigenous tribes, such as the Fishing Festival (Komolis) and Harvest Festival (Ilisin) of the Amis tribe, Ear-Shooting Ceremony (Malahodaigian) of the Bunun tribe and Dwarf Ceremony (Pastaay) of the Saisiat tribe. During the festival, Hafay is mostly tied in sheaves or being processed to liquor (similar to sake) or desserts. See: https://www.fondazioneslowfood.com/en/ark-of-taste-slow-food/hafay-millet/ (retrieved 9/30/2021).

varieties introduced by Han people after coming into contact with them (Huang 1957 [1722]: 95).

This first major change, occurring in the seventeenth and eighteenth centuries, was primarily due to the new regime's policy and a significant increase in the immigrant population. Although there are no accurate statistics, the Han people in Taiwan already outnumbered the aboriginal Taiwanese roughly in the early Qing Dynasty in the seventeenth century.[5] Specifically, the number of Han farmers increased rapidly, and they acquired aboriginal land in various ways for agricultural use. For example, they rented or bought lands from aboriginal tribes with guns, salt, rice, and other objects traded from China or even western people. To increase food supplies for troops, the governments also required the aboriginal Taiwanese to plant rice instead, thus changing the agricultural landscape of Taiwan radically.

During the second significant historical shift, following Japan's colonial rule, new technologies took on a more critical role. After Japan began to colonize Taiwan, many Japanese entered Taiwan, but they were far fewer in number than the Taiwanese people, and most of them were concentrated in urban areas. During this period, the universally planted rice in Taiwan was changed from Indica rice into Japonica rice. Until now, the Japonica rice remains the dominant rice variety in Taiwan. For example, in 2020, Taiwan produced 1,284,970 tons of Japonica rice, 7223 tons of hard indica rice, and 36,581 tons of soft indica rice.[6] As for upland rice, the earliest variety grown in Taiwan, a handful of indigenous groups and scholars have been growing it in small amounts to prevent it from disappearing. Moreover, the quality standard for Japanese rice prevailed in Taiwan, having a far-reaching impact on the development of Taiwan's rice culture. This significant change was primarily due to the introduction of new breeding techniques by the Japanese, irrigation projects, the development of the fertilizer industry, and other infrastructure construction.

[5] According to an official survey by the Taiwan Government-General, which was the highest authority of the Japanese colonial government, in 1926, 88.4% of the total population in Taiwan was Han Chinese. Among those Han, 83.1% were from Fujian and 15.6% from Guangdong.

[6] Agricultural Statistics, Council of Agriculture, Taiwan: https://agrstat.coa.gov.tw/sdweb/public/inquiry/InquireAdvance.aspx (retrieved 3/30/2021).

I now turn to the historical processes accompanying the two major changes just outlined and analyse how immigration and colonization under different conditions substantially changed the staple food of a region.

From Javanica to Indica

Current archaeological studies show that rice was initially planted in Taiwan as early as 4800 years ago, and some rice varieties had the characteristics of Javanica rice (Tsang et al. 2006: 308–309). This may be because, with the emergence of sea routes to Taiwan, prehistoric Austronesian aborigines immigrated to Taiwan and introduced rice varieties of Austronesian regions to this island. The earliest official report on Taiwan written in 1603 pointed out that the long-grain rice grown by the aborigines in southern Taiwan might be a rice variety commonly consumed by all tribes of aborigines in the early seventeenth century (Chen 1959 [1603]: 25).

Javanica rice, also known as tropical Japonica rice, is low-yielding, large-grained, short, and round. Javanica rice is not cold-resistant but highly resistant to drought, leanness, and adverse conditions. Javanica rice is mainly distributed in tropical mountainous areas in the Malay Peninsula, Indonesia, Thailand, and the Philippines (Chang 1976). The cultivation characteristics of upland rice in Taiwan's mountainous regions are similar to those of Javanica rice. This may be because the aboriginal Taiwanese carried Javanica rice to Taiwan to satisfy their demand for a staple food and winemaking when they moved eastwards from the South Sea Islands (Chiang 2004).

In terms of rice cultivation, the aboriginal Taiwanese adopted 'shifting cultivation' (or slash-and-burn agriculture) as their main farming method. Upland rice is drought-resistant and suitable to grow in mountainous areas, and it is mainly grown in drylands or mountain slopes in Taiwan. The water consumption of upland rice is merely one-fifth (or less) of paddy rice (Ding 2016). Upland rice plants are usually tall, thick, and sturdy. Nowadays upland rice is still grown in East Taiwan regions, primarily by aboriginal tribes in mountainous villages.

Although the dry grain yield of upland rice is approximately 3000 kg per hectare now, only half that of Indica and Japonica rice, the upland rice is closely associated with the customs and culture of the aborigines. In addition to brewing rice wine, upland rice forms a crucial ingredient in festive foods. For example, the cooked upland rice is named as *Hahah* by the Amis Tribe, and the mochi made by the rice is named as *Durun*. These foods are served only in banquets or weddings.

With the advent of overseas migration to Taiwan, the cultivation of rice changed considerably. In the eyes of the Dutch, the fertile land in the plains of southern Taiwan could be utilized more efficiently and not be merely used for shifting cultivation. This view was proposed in reports given by Dutch officials and missionaries in 1623, 1624, and 1628 (Chiang 1985: 86, 2007: 144). The Dutch missionary George Candidius (1597–1647) observed that the land of the Siraya people could feed 100,000 people, but they only cultivated the portion of land as needed (Candidius 1994: 224).

The VOC encouraged rice cultivation through various incentive measures. For example, they cooperated with Chinese merchants and local leaders; recruited Han people from Fujian and other regions to move to Taiwan; and gave pecuniary rewards, tax preferences, and even land ownership to Han people who were willing to migrate to Taiwan. As a result of the war between the Ming government and the Manchu—who would establish the Qing Dynasty later—many Han people migrated to Taiwan from war-torn areas, thus bringing rice-farming techniques, different varieties of rice, and farming tools to Taiwan. In 1643, the VOC began to levy a rice tax, because Taiwan had attained considerable achievements in rice cultivation (Heyns 2002: 173), showing the obvious advance in rice farming. To improve the efficiency of farming, cows were introduced from China to Taiwan, bearing key responsibilities of cultivation and transportation. Other than cultivating rice and wheat as food, the Dutch also grew sugar cane, ginger, cotton, and other cash crops for export through international trade. All these changes brought new ecological factors to the local environment, gradually transforming the island's ecological landscape.

After the Dutch, Zheng Chenggong led his troops to Taiwan in 1661 and forced the VOC to leave. He recaptured Taiwan, mainly intending to feed his troops and attack the troops of the Qing Dynasty in Mainland China. To obtain sufficient food supplies for his troops, all soldiers under his leadership took part in making space for farmland. Taiwan's agricultural pattern was transformed into intensive agriculture based on household production (Tseng 2007: 21).

In 1683, the troops of the Qing Dynasty entered Taiwan. Due to the lack of food in the coastal regions of southeastern China, the troops encouraged people to immigrate to central and northern Taiwan to convert land for agricultural purposes in the early eighteenth century. At that time, Taiwan already had a considerably large population of Han population, and Taiwan's agricultural landscape was quite different from that of the seventeenth century. Related records show that in the eighteenth century, the Siraya people in southern Taiwan[7] had already begun to adopt intensive farming methods (e.g., paddy field development and rice transplanting) rather than the original shifting cultivation.

Moreover, the farming workforce and gender division of labour also changed. For the Siraya people, field farming was initially women's duty, while hunting was primarily that of men, who only engaged in farming as a minor duty. As Taiwan's Han population increased, many deer farms dedicated to hunting were occupied and encroached by Han people for conversion into farmland, and men also turned to farming instead.

In sum, before the Han people migrated to Taiwan in large numbers, the aboriginal Taiwanese were already growing rice self-sufficiently, but did not pursue high rice yields or develop any commercial pattern. Subsequent to these Han migrations, government policies encouraged them to plant rice and brought intensive rice-farming modes to Taiwan. When the aboriginal Taiwanese witnessed that the Han people's farming techniques were more productive, they were influenced to adopt new farming tools and methods.

[7] Siraya was the largest tribe of plain aborigines in the south Taiwan in the seventeenth and eighteenth centuries.

From Indica to Japonica

Japanese colonialism actively exploited Taiwanese land for their own agricultural gain. The Japanese regarded Taiwan as a crucial base for their southbound strategy, hoping to make it the centre for rice production to fulfil Japan's needs (Kawano Shigeto 1969 [1941]). The Japanese, however, were not used to the harder texture of the Indica rice grown in Taiwan. For example, cold rice is used for sushi, a significant and traditional Japanese dish, but Indica rice is not appropriate for sushi (just think about the cold curry rice) because it became hard when cold.

To meet the need of Japanese and, more importantly, to reach the goal of forming the 'greater East Asia co-prosperity circle of rice' (Fujihara 2012), the Japanese government actively promoted the breeding and cultivation of Japonica rice in Taiwan. Initially, they tried to change the texture of Taiwan's popular Indica variety by breeding it with one preferred by the Japanese—a softer and stickier rice. But this experiment failed. The Japanese then decided to change their approach by refining Japonica varieties for Taiwan's subtropical conditions and succeeded with this effort in 1922. Breeding was thus the first step in the Japanizing of Taiwanese rice varieties. In 1926, a new variety that grew well in Taiwanese soil was dubbed '*Penglai* rice'. Penglai used to refer to the 'vastness of obscurity' of an 'overseas immortal mountain' (Lu 2019: 129), and the name 'Penglai rice' was given by Izawa Takio (1869–1949), the Japanese governor in Taiwan. By tracing the origins and cultural implications of Penglai, Lu (2019) argued that 'Penglai' was articulated by the Japanese inhabitants of Taiwan as a means of intensifying and naturalizing the cultural connections between Taiwan and Japan. The name 'Penglai Rice' thus aimed to convey both cultural proximity and economic benefits.

However, breeding was just one aspect of rice cultivation. The widespread cultivation of the Penglai Japonica signalled changes in Taiwan's overall agricultural system. For one thing, Japonica rice needs more fertilizer, and its shorter and rounder grain means that different milling machines are needed or else the grains may break. This meant the promotion of Penglai rice represented large-scale alterations in cultivation techniques, irrigation systems, fertilizer manufacture and supply, and the production of milling machines. Previous research has shown the impact on the fertilizer supply chain. In the 1930s, Japan built the largest hydro-electric power plant in East Asia along the Yalu River on the Korean Peninsula, which supplied the Japanese Chisso Corporation with power for the production of chemical fertilizers (Lu 2019). This made Japan the most important country producing nitrogen fertilizer in East Asia. Taiwan's Penglai rice growers thus had access to abundant chemical fertilizers, which in turn facilitated the widespread cultivation of this variety.

New irrigation systems and milling machines, also introduced by the Japanese, created more suitable conditions for the development of Japonica rice in terms of agricultural infrastructure and institutions. Farmers were encouraged to give up familiar techniques of growing Indica rice and learning new techniques. Farmers made the switch to Penglai rice not only under policy pressure, but also because it had a higher yield per unit area. Moreover, in 1930, the wholesale price per 100 kilogrammes of Japonica brown rice in Taipei was ¥12.93, whereas 100 kilogrammes of Indica rice brought ¥10.53. This ratio in prices would be sustained. In 1960, the wholesale price per 100 kilogrammes of Japonica brown rice was NT$488.25, while Indica rice was NT$481.75 (Chen 1962: 8).

After over one million military personnel and civilians moved to Taiwan from mainland China with the Nationalist government after World War II, although most of them were used to Indica rice, the government continue to encourage the production of Japonica rice. The Nationalist government signed a treaty with Japan to trade Penglai rice in exchange for chemical fertilizer in 1953—an exchange that continued until 1969 (Lu 2019), before the end of diplomatic relations between the Republic of China and Japan in 1972. The establishment of this trade in fertilizer created the need for more Penglai rice. The government also exported Japonica rice to Japan and Korea, to earn foreign exchange and gain more taxes (Taiwan People's Voice News, 1952, Central Agency, 1957). In other words, in the post-war period, Taiwan's government encouraged the production of Japonica rice mainly because of its value in international markets, particularly in East Asia regions. Indica rice, produced more cheaply in Southeast Asian countries such as Thailand, would generate no profit through exports.

The economic value of Penglai rice generated symbolic value for it in Taiwanese society. Xu Chin-zhong (1907–1996), chief agricultural officer during the post-war period (1949–1954), claimed that Penglai rice contributed greatly to Taiwan's agriculture sector thanks to the significant income it generated (Wang 1950). The 'Father of Penglai Rice', Iso Eikichi, was one of the few Japanese who remained in Taiwan after World War II. He worked as a consultant in the Department of Agriculture until his retirement in 1957, and the Taiwan Provincial Council decided to provide him free Penglai rice for the rest of his life,[8] showing him high respect and honour, even if he was from the opposing side during the war. From policies that encouraged production to general praise for the Japonica Penglai rice, this variety had clearly and quickly taken root in Taiwan after the war.

Since the 1960s the preference of Taiwanese people had started to change from the dry loose Indica rice to soft, sticky Japonica varieties. An example is the establishment of 'vinegar-rice shops' and 'Han-Japanese restaurants' in the post-war decades. These shops provided the so-called Japanese dishes in Taiwanese style. That is, Japanese food elements were added to local dishes, while the amount was larger and the price was cheaper than Japanese dishes. Sushi was thus sold as 'vinegar-rice,' and Japonica rice was widely used in these shops. The shift of rice preferences did not occur at once; there were several stages in this transition. The first sectors to make this change were military personnel, civil servants, and teachers since their rice supply came from government rations. Hence, any Penglai rice that was not exported was allocated to central government officials, whereas most ordinary farmer households continued to consume Indica rice. For those used to Indica rice, Japonica rice seemed 'too sticky' and some complained it had 'an awful smell' though it was difficult to describe the smell clearly.[9] But eventually, with the increasing production of Japonica rice in Taiwan, more and more farmer households started to plant and eat Japonica rice. In 1961 the production of Japonica rice stood at 1,303,906 tons and Indica rice was only 647,628 tons; and

[8] *Newsletter of Taiwan Provincial Temporary Assembly* 10 (1957/6/2) (002-03-01OA-10-5-3-04-01547) P.10584.

[9] Field note of author, 2019/5/1, 2019/10/15.

in 1971, the Japonica production reached 1,777,703 tons while Indica had decreased to 476,166 tons. In 1981 the production of Indica rice had dropped to less than 200,000 tons (Council of Agriculture 2012: 25–26). This statistic shows how rapidly the production of this variety had decreased, and by the 1970s a steep decline in Indica's popularity had followed.

Japonica rice was served more widely in restaurants and on public occasions from the 1970s. For example, in 1973, the councillor of Taiwan Province, Lin Gengqin, asked government officials in a general meeting of the Provincial Council why the dining cars of trains provided Indica rice instead of Japonica rice:

> I once had a meal with my friend in the dining car of a train, where I found they served Indica rice. This will mislead tourists [into thinking] we only can offer Indica rice, when actually we also have lots of Japonica rice … it will make a bad impression on tourists and is an embarrassment.[10]

The officials agreed with this complaint, replying that they would make sure that Japonica rice is offered in dining cars. This confirms that Japonica rice was seen to have a better texture and higher value among the people's representatives as well as government officials (Chung 2017). In addition, since the crop yield was increased in the 1970s so the supply could meet market demand. The definition of 'good quality' transferred from 'high yield' to 'good taste'.

The definition of 'good quality' changed with the transformation of social conditions. During a time when rice supply is insufficient to meet the market demand, crop yield is the most important quality. Since the Qing Dynasty when rice became the most important staple food in Taiwan, farmers would choose high-yielding and pest-resistant rice varieties for local cultivation in accordance with geographical differences (Tsai 2009). Crop yield remained the most important factor for quality determination up till the early post-war era.

---

[10] *Newsletter of Taiwan Provincial Council* 29 (1973/9/17) (003-05-02OA-29-6-7-00-01514) P. 1142.

The standard of 'good rice' started to change in the 1960s. In 1965, the Taiwan Central Bureau of Standards announced the export standards for Japonica white rice and Indica white rice, dividing the white rice into three grades according to the whiteness of the post-milling rice and other qualities. However, the grading did not include palatability and other important factors. The current evaluation standards of the rice quality were established in the 1970s with the subsidy issued by the Sino-American Joint Commission on Rural Reconstruction, but the influence of Japan in quality evaluation became significant in the 1980s.

It should be noted that it had been common to take sweet potato as a substitution of rice by the Han Chinese in Taiwan since the Qing Dynasty. The Han immigrants tended to export sugar, rice, and tea as cash crops and ate food of less economic value in their daily life (Tseng 2006: 185–186; Chen 2020: 89–93). Until the 1960s, sweet potato was widely grown in Taiwan because it could be planted easily and produced large yields. Thus, for Taiwanese peasant households, it was common to grow or buy sweet potato as the supplement of rice, while rice could be sold for a better price. Sweet potato could be cooked with rice or cut into strips to be dried under the sun for later consumption, and dried sweet potato was more popular because it could be preserved for a longer time. It also explains why crop yield was considered as the most important quality of rice.

Nevertheless, the improvement of eating quality became the main consideration of agricultural policies with the further increase of native rice yields, and the Penglai rice gained more popularity in the market. From rice-breeding and rice storage to palatability evaluation, Japanese methods were taken by the governments (Hong 2011; Wang et al. 2014). For example, in 1984, based on the Japanese palatability evaluation of cooked rice, the preliminary set of palatability evaluation methods for laboratory use was developed by the government to breed rice varieties more suitable for the preferences of Taiwanese people (Wang et al. 2014: 58). Sensory testing was also used as a supplementary method of evaluation for grading Japonica rice from the 1980s onwards (Kuo et al. 1987: 239).

Until recently, the preference for rice with a soft and sticky texture is dominant in Taiwan. For example, in the official Champion Rice Competition, the winners are often the varieties of Koshihikari, which is considered the best rice variety in Japan. Even well-known varieties created in Taiwan, such as Taiken 9 rice and Tainan 16 rice, are referred to as 'Taiwanese Koshihikari'. These varieties are popular not only in rice competitions but also in the rice market.

Although the preference for the texture of Japonica rice has been established since the 1980s, Taiwanese rice breeders and consumers have started to chase more local varieties and flavours that reflect the tastes of Taiwanese society, weakening the influence from Japan. In line with the changing definitions of 'Taiwanese cuisine', the preferences vary accordingly (Chen 2020). Whereas 'Taiwanese cuisine' had been identified as 'food of the colony' under the Japanese colonialism, it referred to 'a regional Chinese cuisine' after World War II. Nevertheless, with political liberalization and an increasing emphasis on the subjectivity of Taiwanese culture, Taiwanese cuisine came to occupy a distinctive category. Nowadays, it is commonly argued that Taiwanese cuisine comprises indigenous flavours like aboriginal, Haklo, Hakka dishes, various Chinese regional dishes, and some Japanese ingredients, and thus it can serve as a national symbol of Taiwan. In the context, some steps of de-colonization were taken. For example, the Council of Agriculture reclassified Penglai rice as Japonica rice in 1996 because it had been named by the Japanese colonial government. Moreover, breeders have become consumed with producing rice varieties that have 'Taiwanese characteristics'.

Breeding Rice with 'Taiwanese Characteristics'

With the definition of 'good quality' rice changed from production yield to taste, quite a few Taiwanese breeding experts have been busy cultivating new varieties adapted to the environment and consumer preferences in Taiwan.[11] Improving flavour, rather than increasing yields, would be

[11] In the second half of the 1970s, rice yields increased but consumption continued to decline. In 1976, yields reached 2.71 million tons, which led to storage problems among Taiwan's agri-food institutions (Liang 2008). In addition, rice consumption continued to decrease each year, with the average rice supply per person falling from 134 kilogrammes in 1971 to 76 kilogrammes in 1986, and even further to less than 50 kilogrammes in 2002. See: Agricultural Statistics: http://agrstat.coa.gov.tw/sdweb/public/book/Book.aspx.

imperative from the 1980s onwards. After the goal of rice cultivation changed, breeding experts began exploring what types of rice were popular among consumers and the characteristics that allowed them to thrive on Taiwan's farms. Taiken 9 rice remains a success story to the present day. Taiken 9 was developed in 1993. It has since then become the standard for excellence in Taiwan. Although its yield is not as high, the main difference between Taiken 9 and other Japonica varieties is that its texture and flavour are closer to the prized Japanese Koshihikari.

A rather different Taiwanese interest has been aromatic rice, varieties popular in South and Southeast Asia such as Thai jasmine rice and Indian Basmati rice, both of which have a rich aroma. Typically, Japanese rice carries no aroma at all and by Japanese standards, rice should be pure and without aroma, since it distracts from the original flavour of the grain.[12] The breeding of aromatic rice varieties started in the early 1980s, but farmers did not like to grow it due to its lower yield, and so it was rarely seen on the market. Not until the change in breeding goals in the late 1980s did aromatic rice gain greater recognition. Since aromatic rice commanded higher prices on the international market, the Council of Agricultural Executive Yuan encouraged farmers to grow it to improve their income levels.[13]

Consumers in South and Southeast Asia have long been fond of aromatic rice, since their cuisines feature spices and sauces that have stronger flavours, so a pure rice flavour is not significant in their food cultures. The preferences of Taiwanese consumers lie in between those of East Asia and Southeast Asia. They accept aromatic rice but do not like overly strong flavours. Some varieties popular in Southeast Asia are thought to smell like disinfectants in Taiwan. Breeding experts have decided that a rice variety needs to have some aroma, but not too much.

Currently, the most popular aromatic rice variety in Taiwan is Tainung 71, which was developed in 1991 but only gained recognition in 2000. Tainung 71, also known as Yichuan aromatic rice, was named after the breeder Yi-Chuan Kuo. It is also a result of crossbreeding Japanese and Taiwanese varieties, specifically, Kinuhikari and Taiken 4. Tainung 71 has

---

[12] *Author's interview with breeding experts.*

[13] *Author's interview with breeding experts.*

the faint aroma of taro when cooked, and since taro is familiar to the Taiwanese, this aroma is well accepted by consumers. After Yichuan aromatic rice gained popularity, dozens of other aromatic varieties were introduced to the market. The rice market in Taiwan was then divided into sectors dealing in aromatic and non-aromatic rice. Besides the taro aroma, popcorn and jasmine aromas are also popular. Chih-Sheng Sheu, who bred Taiken 9, also created Taichung 194, which has the aroma of pandan. These aromatic varieties are markedly different from the Japanese varieties on the market.

Conclusion

The trajectory of rice cultivation and consumption in Taiwan over the past four centuries tells a story about the complex interactions among colonialism, imperialism, immigration, national cuisines, and the world market. It would be tempting to reduce the imposition of Japonica rice varieties on the island to a settler-colonial domination over local food cultures. Yet, as argued throughout this chapter, the picture is more complex, incorporating bio-environmental determinants as well as more contingent cultural and political factors. Since the seventeenth century, with the immigration of Han people and their different means of livelihood; the introduction of cow and various plants by the Dutch; and the significant changes of agricultural methods from the slash-and-burn style to the centralized and intensive style, the colonizers and immigrants transformed Taiwan's ecology and environment. The promotion of rice farming and the forced displacement of aborigines radically reshaped the agricultural scene in the mountainous regions, whereas millet and upland rice disappeared. Moreover, the establishment of modern irrigation system and fertilizer industry further remould the soil, landscape, and rice varieties.

Taiwan's distinctive ecology, with its limited lowland capacity for intensive rice cultivation, meant productivity initially trumped flavour in the choice of rice varieties, as overseas Qing and subsequently Japanese powers administered the island. Colonial rule certainly imposed this measure of 'quality' based on metropolitan standards and the culinary preferences of Han immigrants and Japanese officials thereafter. This hierarchy of taste and economic imperatives was reinforced by the proclamation in Taiwan of the Kuomintang government and the accompanying military-administrative institutions of the new Republic.

But even after this succession of foreign-imposed transformations in Taiwanese rice cultures, there remained a local taste for softer, stickier, and more aromatic grain in contrast to the conventional Japonica rice, which by the 1990s resulted in the creation of the Tainung 71 variety with 'Taiwanese characteristics'. Here, the rise of a hybrid (literally) or creole staple reflects the resilience of autochthonous preferences and their material manifestation in the cross-breeding of varieties.

Chinese overseas settlement and colonization of Taiwan has not fully erased indigenous foodways, but rather adopted and adapted them to wider international market preferences. In contemporary Taiwan, 'searching for the genuine Taiwanese taste/foodway' has been the main theme in the recent decade along with the changing international status of Taiwan. 'Taiwanese local food' is always highlighted as a distinctive offer in restaurant menus, and fusions of French/Nordic cuisine with Taiwanese styles have emerged. The revival of aboriginal food has been remarkable in other forms: millet, djulis (*Chenopodium formosanum*), maqaw (a native tree in Taiwan; its seed is used as a spice/seasoning) which had once disappeared from the table become popular and commercialized planted with the emphasis on 'Taiwanese taste'. Indeed, as Tomonori Sugimoto (2018a) has recently illustrated, there has been a revalorization of indigenous Austronesian foodstuffs and preparation methods, particularly of 'wild plants' (*dateng* in Pangcah/Amis dialects, and *yecai* in Mandarin) among the wider Taiwanese population—even if this comes with significant political-economical and cultural sacrifices.

Although the aboriginal population is tiny, their Austronesian culture plays a crucial role in building up Taiwan as a distinct nation in discourse (Chen 2013: 323).

The case of rice reveals the native efforts at branding Taiwanese as a staple food in spite of the long-lasting influences of settler colonialism. As suggested in this chapter, the changing norms of rice production and consumption in Taiwan can add to ongoing debates regarding the tensions between the 'logic of extermination' and the 'logic of incorporation' in diverse settler-colonial experiences. The Taiwanese case elucidated here shows a staple like rice playing a critical role in what Sugimoto (2018b) has called a process of settler-colonial incorporation and inheritance among parts of the Taiwanese population. The 'self-indigenization' of Taiwanese descendants of pre-1949 Han migrants to the island—the *benshengren* ('those who are native to the province')—which Sugimoto identifies in the cultural and political spheres of contemporary Taiwan is arguably also present in the championing of local rice varieties like Tainung 71, which help to underscore Taiwan's cultural differentiation from the Chinese mainland. Settler-colonial foodways, as we have seen in this chapter and in other parts of this volume, can help to distinguish between forms of dominant assimilation, which take Indigenes as their main subject, in contrast to expressions of incorporation and inheritance, which aim to integrate often stylized and hybrid food legacies into the settler-colonial 'entity'.

Bibliography

Candidius, George, and Chun-zhong Ye (trans.). 1994. Hezhi chuchi de Siraya Pingpuzu. [Plains Tribe Siraya in the Early Dutch-Rule Period]. *Taiwan Fengwu [Taiwan Folkways]* 44 (3): 193–228. [in Mandarin].

Chang, Te-tzu. 1976. The Origin, Evolution, Cultivation, Dissemination and Diversification of Asian and African Rices. *Euphytica* 25: 435–441.

Chen, Di. 1603. Dong Fan Ji [Writing Eastern Barbarians]. In *Minghai Zhengyen [Given Words on Fujian and Beyond]*, ed. Sheng, You-zong, Taiwan wenxian congkan [Taiwan Documentary Collection] Title 56. [in Mandarin].

Chen, Cheng-siang. 1962. Taiwan zhi daomi jinji jiqi yanbian. [Taiwan's Rice Economy and Its Changes]. *Journal of the Agricultural Association of China* 39: 1–14. [in Mandarin].

Chen, Yujen. 2013. Ethnic Politics in the Framing of National Cuisine: State Banquets and the Proliferation of Ethnic Cuisine in Taiwan. *Food, Culture and Society* 14 (3, Sept.): 315–333.

———. 2020. *'Taiwancai' de wenhuashi: shiwu xiaofei zhong de guojia tixian. [The Cultural History of 'Taiwanese Cuisine': Embodying Nation in Food Consumption]*. Taipei: Linking. [in Mandarin].

Chiang, Shu-sheng (trans.). 1985. Xiao long cheng ji. [Notes of Xiaolong Tribe] *Taiwan Fengwu [Taiwan Folkways]* 35 (4): 80–87. [in Mandarin].

Chiang, Rui-gong. 2004. Ludao Jichi Zhangwang. [Upland Rice and Its Future] *Nongye Zhijie [Agriculture World]* 248: 58–62. [in Mandarin].

Chiang, Shu-sheng (trans.) Taiwan Historica (ed.). 2007. *Helan Taiwan Zhangguan zhi badaweiya zhongdu shuxinji 1 1622–1626 [Letters from Dutch Governors at Taiwan to the Dutch Governors at Batavia 1622–1626]*. Taipei: Nantian Publishing. [in Mandarin].

Chung, Yi-Ting. 2017. Taiwan mi, Ribenwei: Guanyu hao de pinzhi jiangou. [Taiwanese Rice, Japanese Taste: The Construction of 'Good' Quality]. *Journal of Chinese Dietary Culture* 13 (2): 13–50. [in Mandarin].

Council of Agriculture. 2012. *Taiwan Food Statistics*. Taipei: Council of Agriculture.

Ding, Wen-yen. 2016. Ludao zaipei guanli moshi. [The Model of Upland Rice Planting]. *Agriculture Newsletter of Taidong* 97: 13–16. [in Mandarin].

Fujihara, Tatsushi. 2012. *Ine no Daitōa Kyōeiken: teikoku nihon no midori no kakumei [The Greater East Asia Co-Prosperity Circle of Rice: Imperial Japanese Green Revolution]*. Tokyo: Yoshikawa Kōbunkan. [in Japanese].

Heyns, Pol, and Weizhong Zheng (trans.). 2002. *Helan shi dai Taiwan de jing ji, tu di yu shui wu [Economy, Land Rights and Taxation in Dutch Formosa]*. Taipei: Bozhongzhe. [in Mandarin].

Hirano, Katsuya, Veracini Lorenzo, and Toulouse-Antonin Roy. 2018. Vanishing Natives and Taiwan's Settler-Colonial Unconsciousness. *Critical Asian Studies* 50 (2): 196–218.

Hong, Mei zhu. 2011. Daomi pinzhi yu shiwei pingjian shizuo [Rice Quality and Practice of Sensory Evaluation]. http://www.tcavs.tc.edu.tw/upload/1001206141556.pdf. [in Mandarin].

Huang, Shu-Ching. 1957 [1722]. Taihai shichalu. *[Writings on Taiwan Trip]*. Taiwan wenxian congkan [Taiwan Documentary Collection] Title 4. [in Mandarin].

Kawano, Shigeto, and Lin Ying-yen (trans.). 1969 [1941].*Riju shitai Taiwan migu jingjilun [On Taiwan's Rice Economy During the Japanese Ruling Period]*, Taipei: Taiwan Bank Economy. [in Mandarin].

Kuo, Yih-chuan, Ching Liu, and Song Sing. 1987. Mifan wulixing yu guanneng pinchang zhi guanxi. [The Relationship between Human Panel Testing Score and Physical Properties of Cooked Rice]. *Journal of Agricultural Research of China* 36 (3): 239–246. [in Mandarin].

Li, Yi-yuan. 1982. *Taiwan tuzhu minzu de shehui yu wenhua [Society and Culture of Taiwanese Indigenous]*. Taipei: Linking. [in Mandarin].

Liang, Bang-kuen. 2008. Nongye Shiwu Zhaidihua: Taidongxian Chishangxiang Mishichanye de tantou. [Localization of Agro-food Exploration of Rice Industry within Chin Shang Township in Taitung County]. *Journal of Geographical Science* 53: 85–117. [in Mandarin].

Lu, Shao-li. 2019. Savage World, Immortal Island: The Colonial Gaze and Colonial Taste of the Penglai Rice. In *Positioning Taiwan in a Global Context: Being and Becoming*, ed. Bi-yu Chang and Pei-yin Lin, 128–143. London: Routledge.

Tomonori, Sugimoto. 2018a. 'Someone Else's Land is Our Garden!': Risky Labor in Taipei's Indigenous Food Boom. *Gastronomica* 18 (2): 46–58.

———. 2018b. Settler Colonial Incorporation and Inheritance: Historical Sciences, Indigeneity, and Settler Narratives in Post-WWII Taiwan. *Settler Colonial Studies* 8 (3): 283–297.

Tsai, Cheng-hao. 2009. *Taiwan daozuo jishu bianqian zhi yanjiu [Research of the Taiwan Rice Technology Evolution (Prehistoric to 1950)]*. PhD thesis, National Taiwan Normal University, Taipei. [in Mandarin].

Tsai, Cheng-hao, and Yun-ping Yang. 2004. *Taiwan Fanshu Wenhuazhi [The Trace of Taiwan Sweet Potato]*. Taipei: Guoshi Publishing. [in Mandarin].

Tsang, Cheng-hwa, Li Kuang-ti, and Zgeng-yi Chu. 2006. *Xianmin Luji [Tracks of the Ancestors]*. Tainan: Tainan County Government. [in Mandarin].

Tseng, Pin-tsang. 2006. *Cong tianqi dao canzhuo: Qingdai Taiwan hanren de nongye shengchan yu shiwu xiaofei [From Farm to Table: The Agricultural Production and Food Consumption of the Taiwanese Han People in Qing Dynasty]*. PhD Thesis, National Taiwan University, Taipei. [in Mandarin].

———. 2007. Wujing yu zenze: Hezhi yu mingzheng shichi Taiwan de nongye fazhang yu huanjing gaizao. [The Development of Agriculture and the Environmental Reshape in Taiwan during the Dutch Period and Cheng Chen-kung Period] *Bulletin of Academia Historica* 14: 1–38. [in Mandarin].

Wang, Guo-bing. 1950. Penglaimi de youlai [The Origin of Penglai Rice]. *Central Daily News*, September 12, p. 8. [in Mandarin].

Wang, Ying-zeng. 1962 [1764]. *Chongxiu fengshan Xianzhi [Gazetteer of Fengshan County]*. [in Mandarin].

Wang, Bo-rong, Hong Mei-zhu, Yang Jia-ling, and Jian-feng Li. 2014. Zui daodi de mizhi yanjiu shi [Research History of Rice Qualities]. In *Proceedings of the 30th anniversary research and extension achievements by relocated Taichung District Agricultural Research and Extension station*. Changhua, COA. [in Mandarin].

5

'Like the Papacy of Mexican Cuisine': *Mayoras* and Traditional Foods in Contemporary Mexico

Claudia Prieto-Piastro and Alejandro Colás

Introduction

The figure of the *mayora*—the wife or concubine of the Spanish American Hacienda owner and therefore principal cook in the Estate household during the colonial period—has over the past few decades been recovered, revalorised, and promoted in Mexico as a purveyor of ancestral culinary knowledge among First Nations (*pueblos originarios*). Contemporary *mayoras* are deemed to be and generally self-represent as (overwhelmingly female) cooks that recover, sustain, and disseminate the 'intangible heritage' of traditional Mexican cuisine. In the words of the submission for UNESCO recognition: 'Traditional Mexican cuisine is a comprehensive cultural model comprising farming, ritual practices, age-old skills,

C. Prieto-Piastro (✉)
Brunel University London Pathway College, London, UK
e-mail: claudia.prieto-piastro@brunel.ac.uk

A. Colás
Department of Politics, Birkbeck, University of London, London, UK
e-mail: a.colas@bbk.ac.uk

R. Ranta et al. (eds.), *'Going Native?'*, Food and Identity in a Globalising World,
https://doi.org/10.1007/978-3-030-96268-5_5

culinary techniques and ancestral community customs and manners' (UNESCO 2010). *Mayoras* include *cocineras tradicionales* ('traditional cooks') whose 'knowledge and techniques', according to the Conservatory of Mexican Gastronomic Culture (CCGM), 'express community identity, reinforce social bonds, and build stronger local, regional and national identities [while they] also underline the importance of traditional cuisine as a means of sustainable development'.

There are, however, several degrees of ambiguity in such statements. Not all traditional cooks are *mayoras* (the CCGM does not employ this term), but the latter do seem to all count as traditional cooks. Indigeneity—that is, having as a mother tongue one of the rich varieties of pre-Hispanic languages—is one way of distinguishing traditional cooks from *mayoras*. Cooking primarily for, and within a First Nation community, without a commercial or profit motive appears to be a further characteristic of traditional cooks in contrast to *mayoras*. Neither of these groups of female cooks are ever considered 'chefs', yet their ingredients, recipes, techniques, and preparations have increasingly found their way into *mestizo* repertoires reaching into fine dining, or the innovative 'cocina de autor/a'.

We seek in this chapter to probe such classificatory ambivalences and explore whether and how they reflect a contradictory set of processes whereby gender, ethnicity, indigenous identity, and popular culture are classified, regulated, and commodified in order to deliver a codified notion of 'traditional' Mexican cuisine. For while it is incontrovertible that Mesoamerican ingredients like maize, beans, and chillies and their associated horticultural, gastronomic, and cosmological practices and rituals have for millennia formed the backbone of many Mexican cuisines, this is also the case for multiple and much more recent post-Columbian foodstuffs, techniques, and culinary traditions. A convenient way of reconciling these creative tensions is to fall back on the notions of *mestizaje*, transculturation, or hybridity as an inevitable outcome of the colonial encounter and the subsequent post-colonial admixture of peoples and cultures. But we argue in this chapter that doing so underplays the real, unresolved ambivalence in contemporary invocations of 'traditional' Mexican cuisine, where competing pressures to both acknowledge and safeguard native, pre-Hispanic culinary cultures while also promoting and valorising such heritage for regional, national, and indeed global markets today render the very notion of 'traditional' cuisine inconsistent and probably unsustainable.

As Raúl Matta's illuminating study indicates, traditional cooks have emerged out of two processes in Mexico: 'the commodification of culture and the neoliberal management of cultural heritage' (2019, 212). For Matta, *cocineras tradicionales* are marketed to the outside world by local, regional, and national agencies as bearers of indigenous cultural authenticity, yet they are also encouraged to become 'enterprising subjects who are able to cope despite adversity and lift themselves out of poverty' (Matta 2019, 227). Our own research chimes with these conclusions, but we aim to establish in what follows that there is a genuine material contradiction at play in attempts at reconciling the value of indigenous Mexican cuisines with the projection of *mayoras* as traditional cooks today. There is, we argue, both a socio-economic tension in deploying neoliberal conceptions of 'ethnopreneurialism' to combat the collective marginalisation of indigenous communities caused by centuries of colonial racism and a socio-cultural impasse when seeking to reconstruct and patent traditional Mexican cuisine under post-colonial circumstances where what counts as 'native' gastronomy has become so indeterminate.

We develop our argument in three steps. The following section outlines the primary research we conducted in the course of 2020. This entailed four open-ended interviews with traditional cooks from different regions in Mexico, as well as Mtra Gloria López Morales, President of the *Conservatorio de la Cultura Gastronómica Mexicana*. The complex, changing and in places, conflicting (self-)representations of traditional Mexican cuisine come across in the subjective reflections among *mayoras* and *cocineras tradicionales* on their practice and role in Mexican gastronomy. A closing section moves to a higher level of abstraction and weighs out the benefits and shortcomings of invoking *mestizaje* or transculturation as a resolution of the socio-cultural aporia just mentioned (Miller 2004). Plainly the Spanish conquest and subsequent settlement of peoples from all corners of the world (whether free or enslaved) led to the emergence of distinctive Mexican foodways which, in their uneven combination, have delivered what in other colonial settings might be labelled a creole and, in Mexico, a mestizo cuisine.[1] But, as we will see, there is a danger of erasing the unequal process and hierarchal outcomes of the Columbian Exchange when invoking *mestizaje*, leading to what Pablo Yankelevich (2012) in another context labels the 'unequal and exclusionary *mestizaje*' which grew out of the post-revolutionary Republic as a signifier of national identity. Rather than leading to a national 'melting pot' of identities, the Mexican cuisines that emerged throughout the colonial period and after Independence display a complex variety of repertoires which—like most other gastronomic cultures—are cut through with class, gender, regional, urban/rural, and of course ethnic cleavages. From this perspective, the *mayora* becomes less a vessel of ancestral continuity and more a subject of contested social identity. We explore towards the end of the chapter some of the comparative and conceptual threads running across this edited volume by considering questions surrounding the historical evolution of the *mayora*, the ambivalences this figure carries with regard to both the legacies of First Nation and settler-colonial heritage, and the implications of all this for a more international perspective on how indigenous or traditional foodways have been reconstructed or reverse-engineered as part of the reinterpretation of contemporary post-colonial cuisines.

Cocineras Tradicionales: Vessels of Collective Ancestral Knowledge

Cocineras tradicionales are often described as female cooks with indigenous ancestry who oversee the preservation and transmission of traditional cooking methods and recipes. The term encompasses a range of women with diverse social as well as cultural backgrounds. Furthermore,

[1] We adopt here the more capacious notion of 'creole' as denoting ethnic admixture emerging in settler-colonial societies, rather than the narrower, etymological conception of 'criollo' as descendants of Europeans born in the New World. It is in this regard that 'mestizo' in Mexico is an equivalent to 'creole' elsewhere and, gastronomically, close to what Donna Gabaccia (2000) understands by 'colonial creoles' across the Americas, or Richard Wilk the process of 'creolisation' in Belize: 'hardly a smooth blending process [but a] compound of appropriation and resistance, full of ambivalence and ambiguity' (Wilk 2006, 109).

it displaces ethnicity in favour of a nationalist understanding of Mexican food culture in which some regional and local practices are celebrated whilst others are overlooked. The use and popularisation of the term 'cocinera tradicional' is the result of complex processes of heritagisation in culinary practices (Bessière and Tibère 2011; Brulotte and Di Giovine 2014; Matta 2019) which have replaced the original designations given to indigenous cooks such as *comidera* and *mayora* and even supplanted labels used by cooks to describe themselves.

The origins of the term *mayora* and its uses in the Mexican food culture are obscure, thereby accounting for the lack of consensus on who can or should be labelled as one. Originally, the category simply designated the wife of a *Mayor*, a title derived from Spanish nobility. It is unclear how this description came into use when referring to a female Mexican cook, but it is still common to find cooks employing it to indicate their status in a kitchen. The *Diccionario Enciclopédico de la Gastronomía Mexicana* (Zurita 2010) does not include an entry for *mayora*[2] or for *cocinera tradicional* but does include one for *comideras*.[3] There is, again, no entry for *mayora* or *cocinera tradicional* in the *Diccionario del Español de México*, but there *comideras* are defined as female street food vendors (El Colegio de Mexico).

Gloria López Morales, president of the CCGM and former civil servant behind the Mexican candidacy for inscription in the UNESCO list, defines *mayoras* as 'skilled rural women that have learned to cook in their communities, but cook inside the structure of commercial cuisines. They spend years working on restaurant chains or small *fondas*, without their skills being valued. They are home cooks, chefs without titles'. This definition is backed up by cultural officials and chefs. Nonetheless, the *cocineras* interviewed for this chapter disagree with these characterisations. According to the *mayoras* themselves, they represent the highest authority

[2] The dictionary does include an entry for *Mayordomía* (Mexican traditional festivity) but does not acknowledge those in charge of its organisation as *mayoras*.

[3] Local term used in Oaxaca to describe women that achieved the highest position in a local kitchen. According to the *Diccionario Enciclopédico de la Gastronomía Mexicana*, a *comidera* is elected by their community and oversees the organisation of regional festivities. The most famous *comidera* is Abigaíl Mendoza, traditional cook from Teotitlan del Valle, Oaxaca, and owner of the storied restaurant Tlalmanalli (Muñoz Zurita 2010).

in a Mexican kitchen, independently of the tools, techniques, and materials employed or the organisation and location of their food preparation and cooking space. In the words of Nelly, a cook from Tabasco, *mayoras* 'are like the papacy of Mexican cuisine. If the *mayora* tells you to boil the broth for two hours, you should boil it for two hours, no questions asked'.

In contrast to the popular terms *comideras* and *mayoras*, the descriptor *cocinera tradicional* is a heritage word constructed around the image of an indigenous woman—a third-generation female cook who acts as the vessel of highly codified communitarian culinary knowledge and who possesses the skills required to utilise traditional Mexican cooking tools such as the *metate* (UNESCO 2010). Gloria López Morales defines *cocineras tradicionales* 'as those women that cook in their community in a traditional kitchen. It is just now that they are starting to go out to the world to demonstrate their skills. Recently, and thanks to the guidance provided by the CCGM, they are opening small food businesses'. Tourism is fundamental for them; the cooks that live in tourist regions are more successful than the others. According to Gloria's definition, a traditional cook is characterised by her knowledge of ancestral cooking methods and her usage of traditional tools and spaces, not by ethnicity or geography, although she stresses the importance of communitarian culinary knowledge. Gloria also acknowledges that there are thousands of women in Mexico that can be labelled as traditional cooks, and she also recognises that they can also be found on the 'other side of the border' (in the United States).

The dossier presenting the Mexican candidacy to the UNESCO establishes that *cocineras* are required to attend the events organised by the CCGM and are responsible for building and decorating their own stands, transporting their cooking tools, as well as sourcing their ingredients (UNESCO 2010, p. 8). In addition to presenting food that is cooked using 'artisanal' methods, the dossier emphasises the importance of aesthetics: the *cocineras* should wear for these events their everyday traditional garments and should make sure the interior and exterior of their stands or houses are decorated in accordance with what visitors expect to see in traditional practices (UNESCO 2010).

The term 'indigenous' remains undefined in the dossier: it is used as a synonym of communitarian, traditional and autochthonous, and the delimitation of the term remains unclear. Therefore, the designation as *cocinera tradicional* not only includes indigenous women but may also refer to any expert in community cuisine, independently of their ethnic background. The ambiguity of the label makes its use controversial since the term fails to recognise the cooks' regional and ethnic diversity as well as their culinary background. In addition, it creates the illusion of an *haute* traditional cuisine—a concept that has become fundamental in Mexico's nationalist heritage discourse (Hryciuk 2019, p. 10)—thereby leaving those women that consider themselves community cooks but not acknowledged as such by the CCGM unable to access similar resources to those that do have a positive relationship with, and are fully recognised by, the institution. The lack of clarity over the cooks' ethnicity provokes anxiety among some of the *cocineras* who see in their *mestizo* background an impediment to their official identification as traditional cooks. All these issues of identity and recognition in turn play into the very material considerations of market share—or at least opportunities—in a thriving indigenous gastronomy and heritage tourism sector (de Sureiman 2017). Together, such questions of indigeneity, official accreditation, and ancestry were raised by our informants as key areas of concern when discussing how pre-Hispanic culinary legacies manifest in traditional Mexican cuisine today, and what role *cocineras tradicionales* play in this unfolding story.

Who Can Be a *Cocinera Tradicional*?

Nelly is one of the cooks that is anxious about the official definition of *cocinera tradicional*. Although one of the most successful and well-known *cocineras*, she feels her *mestizo* background and her higher education do not perfectly fit this label. Despite carving her own culinary path, she perceives the CCGM's indifference towards her as a reflection of her 'dubious' traditional background.

We interviewed Nelly via Zoom in September 2020, just a few days after her restaurant in Comalcalco, Tabasco, was named one of the best 20 restaurants in the world by Besha Rodell, a culinary researcher from the magazine *Food and Wine*. Even though Nelly had been interviewed several times that week, she was quick to answer us and happy to talk for as long as we wanted. For the first half of the interview, she did not mention the recent success of her restaurant, focusing instead on the economic difficulties she and her staff were facing since the beginning of the COVID-19 pandemic.

Nelly studied law at a local university and worked as a lawyer for decades before opening her restaurant. In most of her interviews, Nelly mentions a passion for cooking as well as her family's humble origins. Nelly comes from a middle-class family in Tabasco, and her grandfather took care of her while her mother was working. She liked cooking since childhood, and while not allowed to cook in the family kitchen, she did learn recipes in a theoretical way, slowly building a mental archive of cooking techniques and ingredients. Nelly spent approximately half the interview talking about her childhood: how she grew up playing with her sister in their grandfather's 'rancho' and constantly mentions her rural origins, her knowledge of the land, and her grandfather's fields. Whenever possible, Nelly talks about growing up surrounded by cocoa plantations and of how she learned from her grandfather the correct way of drinking chocolate: 'not too hot and not too cold, not to sweet and not bitter, not to light and not to dense' (Córdoba 2018).

When Nelly talks about other *cocineras tradicionales*, she does so with admiration and respect, but she also mentions that many of them are poor and uneducated; women that had no other skill than cooking and therefore had to use that one asset to survive. She struggles to say it, but Nelly also thinks they should have an indigenous background, and it is therefore evident she feels her story does not fit the conventional narrative. Having a Spanish grandfather makes her a 'mixed-race' *mestiza*, and her career as a lawyer and her socio-economic status also seem to make her stand out from other *cocineras*, at least in her view. Nelly constantly mentions that she has a humble, lower-middle-class origin, as this could compensate for her education and the nationality of her grandfather. The rural stories she tells about her childhood seem to also have the aim of proving an almost innate connection to the land and the native ingredients of Tabasco that could compensate for her lack of ties with the local communities of the region. Nelly's economic struggles are a fundamental part of her tale, while she also downplays the privileges that came with her education and family's socio-economic position. 'I come from a modest, humble family; we are proud of our roots and our traditions', she repeats during our interview.

Nelly does not hide the fact that she sees herself as different from other *cocineras*. She believes she has been able to build a food business from scratch without the help of any external actor, and she also feels she does not respond to the stereotype of the 'poor, indigenous, uneducated women' who generously share their culinary expertise—a body of knowledge that should be inherited from a community, which is not her case. Nelly is both proud and anxious about being an outlier in this respect. She encourages other *cocineras* to put a fair economic value to their work and complains about the lack of support they receive when they cook at gastronomic conferences and international events. Whenever possible, she avoids working for free or without compensation. She feels that denouncing the conditions in which many renowned *cocineras* live is one of her responsibilities and does not shy away from the topic, explaining in detail the hardships *cocineras* endure. Nonetheless, Nelly resents the fact that she is sometimes not invited to the events organised by the CCGM or the Mexican government and believes her ethnic origin and education are the reasons why she is not considered a *cocinera tradicional* by the authorities.

Nelly does not have an indigenous background to prove the 'authentic' roots of her food, nonetheless the name of her restaurant 'Cocina Chontal' refers to the Mayan natives of Tabasco known as the Chontales. Her food is served in pots and plates crafted by local indigenous communities, and the clothes she wears also come from these communities. But her menu is rooted in regional creole specialities such as enchiladas filled with pork leg, beef *chirmol*, and *pejelagarto* (fish) in green salsa, and she seems to prefer talking about her food as regional rather than indigenous.

Although Nelly sees her story as unique, she is not the only *cocinera* that has been able to build a successful food business by utilising the strategic essentialism of the official, culinary heritage narrative. Abigail Mendoza, the owner of the renowned restaurant Tlamanalli in Teotitlan del Valle, Oaxaca, has a similar story. She is a middle-class woman from a family that runs a profitable textile business, with connections outside Mexico, and supported by the male family members. Her fame began in

1993—years before Mexican gastronomy was added to the UNESCO intangible list—when a reporter from the *New York Times* named Tlamanalli a 'top notch table' (O'Neill 1993), praising her moles and fried fish. *Mendoza*'s reputation was further enhanced, as her restaurant featured in an episode of Anthony Bourdain's TV show *Parts Unknown*, following his visit there in 2014.

Like Nelly, Abigail has built a narrative in which her family's economic success hides behind a passion for cooking and a connection to the land—in Abigail's case, the fields where her parents harvested maize (Hryciuk 2019, 11). In contrast to Nelly, Abigail's association to an indigenous community is explicit. Her mother tongue is Zapotec, and she is a third-generation cook—both aspects emphasised in her talks and interviews. Her menu stresses the use of pre-Hispanic Zapotec cooking techniques, ingredients, and tools, even when most of the dishes listed make use of colonial ingredients such as *quesillo*,[4] sugar, rice, and milk.

Both Abigail and Nelly talk about being hard-working, self-made women and downplay any external help they might have received from any institutions. Nelly refers to the influence of her grandfather constantly, but never acknowledges any economic support she received from him. She is visibly uncomfortable when mentioning the 'woman that helped them around the house, like a nanny', and tries to balance this fact by noting her family never had a television (Cordova 2020). Abigail openly talks about the encouragement received from her father, a former mayor of Teotitlan, but excludes from her personal narrative her family's economic success (Hryciuk 2019, 11).

Nonetheless, the success of both women cannot only be attributed to their indisputable determination and knowledge. Their skill in mobilising heritage discourses and culinary tourism, enhanced by the association to UNESCO, has played a key role in their success. It is no coincidence that Nelly's restaurant is only 150 m from the archaeological site of Comalcalco, while Abigail's Tlamanalli is in one of Oaxaca's unmissable tourist destinations.

But the proximity to an archaeological site or a heritage route is not enough to guarantee the economic stability of other *cocineras*: even those

[4] Also known as Oaxaca cheese, it is a string semi-hard cow's cheese similar in texture to mozzarella.

well-known *cocineras* who own businesses are facing difficulties due to the pandemic. Eloy, a *totonaca* traditional cook from Veracruz, told us that from March 2020, the collective of cooks to which he belongs had to look for alternative economic activities to guarantee an income.

Eloy is the only male in a collective named *Mujeres de Humo* (Women of Smoke), the *mayoras* of the *Casa de la Cocina Tradicional Totonca (The House of Traditional Totonaca cuisine)*, located in the *Centro de las Artes Indigenas* (CAI), an institution registered with the UNESCO List of Best Practices for the Safeguarding of Intangible Heritage (2012). In their words, this association increases their heritage value (Mujeres de Humo 2018).

Eloy has been a part of this collective for more than a decade and he is now the right hand of the woman whom he calls the *mayora de mayoras*, Martha Soledad Gomez Atzin. Eloy is in charge of administrative tasks and transporting cooking equipment, but also of cooking and baking together with his mother, who is part of the collective too. Eloy tells us that one of the reasons he was accepted as part of the group was his strength, an important aspect to consider given the heavy weight of many tools used in a traditional Mexican kitchen.

During the interview we conducted with him in October 2020, Eloy focused on the challenges *Mujeres de Humo* have faced during the year, and how their main sources of income have disappeared due to the lack of tourism and support from official institutions. According to Eloy, the income they earned through teaching at the CAI has vanished, and they only receive a symbolic payment from the government of Veracruz:

> *It takes me more than an hour to get from my town to the CAI, and the money that I get from the government does not cover fully my monthly transport. That means that if we go there, we are doing it because we want to preserve our traditions and heritage, but we don't even have enough to pay for the ingredients to teach a class. My mother and I are selling bread early in the morning to the locals, and our classes consist of foraging and learning how to classify the herbs of the region: that is the only thing we can do now. If there are no tourists, there is no income.*

Local and international tourism to the Tajin, an archaeological site located near Papantla and the CAI, began to decline a few years before the pandemic. The almost sudden increase in drug-related violence, as well as the construction of a Maximum-Security Prison in the area that took more than 20 years to complete, has affected the image of the region, as well as the living conditions and the state's spending on culture and heritage. It is therefore unsurprising that Eloy and the rest of the cooks feel neglected.

Although the work that *Mujeres de Humo* carry out at the CAI, in addition to their collaboration with the CCGM, has come to a halt over the past year, they have continued with their independent research work in the region as well as their efforts to teach younger generations their language (totonaca) and food culture. Eloy explains that in the last few years *Mujeres* have informally researched the culinary traditions of the neighbouring towns and gathered recipes and cooking techniques that are traditional to the *totonacas*. However, he regrets being unable to classify, process, and publish the latter due to lack of economic support, and he is concerned their efforts might be lost.

Nonetheless, in 2018 *Mujeres de Humo* published a book with support from the Culture Ministry (Mujeres de Humo 2018) that, according to Eloy, was a success. The book includes contributions from cultural officials praising the group for the work they have done to promote their culinary heritage, as well as poems and vignettes written by the *cocineras* themselves. The book focuses on the presentation of traditional totonaca recipes—most of them well known in the region—and those cooked by the *cocineras* in their homes and at the CAI. But the success Eloy refers to has not translated into economic profit, and the *cocineras* do not consider the support they receive from different institutions is sufficient for them to continue promoting and disseminating their culinary culture. They therefore believe the inclusion of Mexican gastronomy in the UNESCO list has not brought them or their community any tangible benefits.

In contrast to Nelly, the *Mujeres de Humo* impatiently wait for the opportunity to be certified as traditional cooks by the CCGM and the Education Ministry—a programme that has been in the planning for years but has not yet been implemented. Nelly believes the certification will mainly have a negative impact on those *cocineras* like her, who cannot trace back their knowledge to an indigenous community or family. Eloy, on the other hand, believes that the only criteria when deciding who is certified as a traditional cook should focus on the command over traditional cooking tools and the use of an *adobe* (clay) kitchen—both aspects that he feels distinguish traditional cuisine from modern cooking. Eloy states that until this certification is not in place, they will be at a disadvantage, particularly with regard to the government health and hygiene certificates. Eloy explained:

> *According to the CCGM and other institutions, to be considered a traditional cook you need to cook with the tools and in the environment that our ancestors did. And I agree. But the health inspectors do not give us health and hygiene endorsements because they can find dust and ashes in our food. We cook in adobe kitchens, it's impossible to avoid dust and ashes, and maybe the certification will resolve this issue. Maybe it will also allow us to give diplomas to those who train with us.*

Other *cocineras* share Eloy's view of the possible benefits of the certification. One of them is Reyna, a traditional cook from the Estado de Mexico, who hopes that being officially recognised as a *cocinera tradicional*, and the accompanying government support to attend national and international gastronomic events, will improve the modest profits her small food business makes. Reyna lives in the small town of Xonacatlan and sells *mole* paste from home. She is not only a well-known *cocinera*, but a renowned embroidery artist: in 2020 she designed and embroidered the World Boxing Committee belt given to the Mexican boxer Julio Cesar Chavez that same year.

But Reyna does not mention her work as an artisan during our interview in October 2020. Reyna is modest and only talks about her work as a *cocinera* in the framework of her interest in promoting and preserving her *Otomi* culture. She believes it is her duty to teach the new generations of her community their ancestral customs, language, and traditional recipes and spends a considerable amount of time doing so. In recent years, she started selling mole paste to tourists and locals, but the business is not going well:

> *It's not only that tourists are not coming this year, it's that not everyone appreciates the effort that it takes to make mole in a metate. Not everyone can afford it either. I now sell a few different types of mole, I make some in a Molino, so they are cheaper and more people can buy them; and some in the metate, those I label as artisanal.*

Although Reyna acknowledges the economic hardships she is going through this year, she does not focus on them during the interview. She instead details how most of the food she consumes comes directly from the allotments, farms, and rivers in her hometown, and therefore she considers herself lucky. The possibility of growing her own food makes Reyna less dependent on an income for subsistence. Her food is humble, but Reyna is proud of it: she invites us several times to visit her so we can see the allotments and taste her food. At the same time, she remembers episodes from her childhood and youth in which she felt discriminated against because of her indigenous roots, cuisine, and clothes:

> *They bullied me at school because I ate charales* [freshwater whitebait] *and nopales* [cacti] *and because in my house we ate that way. I also remember seeing people laughing about our traditional attire when we went out to the street dressed in them, but I am proud now, I know there is nothing wrong with my food or the way I look, on the contrary, I am trying to save our recipes.*

Among the recipes mentioned during the interview is one that she is particularly proud of—her white mole, also known as bride's mole. Although this recipe includes ingredients first introduced to Mexico by Spanish settlers such as lard and white chocolate, she considers it one of the emblematic *Otomi* recipes and one that she is well known for. But when looking at the origins of this mole and the perception many Mexicans have about it, the *Otomi* roots of the dish are ignored and instead is recognised as a dish from the neighbouring state of Puebla.

For Reyna, as for Eloy, the political borders of Mexico's states are only important when they refer to institutions and resources—they are inconsequential when it comes to establishing cultural divisions between the First Nations of Mexico. These are the ethno-linguistic and gastronomic divisions that are meaningful for Reyna, and therefore her aim is to preserve the culture of her own indigenous group with roots in the Estado de Mexico but also in Puebla, as her *mole blanco* recipe shows. Both Reyna and Eloy point out several times during the interviews that their dishes are not regional or representative of the gastronomy of their state, but instead highlight the indigenous origin of the recipes.

Currently, Reyna acts as the lead between the *cocineras* of Mexico state and the CCGM, making her the representative of *cocineras* belonging to three indigenous groups. She does not seem bothered by this. They live nearby and she knows the other *cocineras* as well as her culinary traditions. But Reyna mentions she feels they have less recognition than other *cocineras* from places such as Michoacán, who attend events accompanied by an officer from the Culture Ministry and receive more support from them.

In sum, the geographical and political division that authorities have made of Mexican cuisine does not reflect the ethnic diversity and location of the First Nations, making the certification of *cocineras* even more complicated. It is still unclear which criteria will be used to evaluate and certify them, but 'testing' their knowledge on regional cuisine could only contribute to the weakening of indigenous cultures. Moreover, by replacing commonplace terminology like 'mayora' and 'comidera', with the umbrella term 'cocinera tradicional', official institutions further contribute towards the erasure of ethnic diversity.

Mayoras: Contested Identities, Ambivalent Figures

The challenges in reconciling individual creativity, communal identity, market pressures, and institutional incentives, as testified above by the purveyors of traditional Mexican cuisine, are not unique to that country. Anywhere gastronomy is valorised as a cultural asset—be in the shape of *terroir* or in the form of heritage food festivals—there is a push on the one hand towards branding and product differentiation directed at the paying customer and the pull on the other hand of a more subjective, emic conception of culinary traditions expressed in practices beyond commercial transactions (Brulotte and Di Giovine 2014). There is no question that in Mexico, the experience of many First Nation communities continues to be one of racist marginalisation, exclusion, and oppression. Yet it is difficult, as we have thus far seen, to identify a recognisable settler food culture to accompany such patterns of domination (aside from perhaps exceptions like *cabrito* from Monterrey, associated with the ranchero frontier). It is also problematic to pin-point examples of pristine indigenous cuisines that do not in some way incorporate post-Colombian ingredients (including, for instance, Reyna's iconic Otomi *mole blanco*).

Instead, the highly diverse repertoires available across the country deliver something closer to an overlapping mosaic of culinary cultures rather than a rigid distinction between settler and native foodways. The hierarchical *mestizaje* of recipes, ingredients, techniques, and diets characteristic of Mexico since the Spanish conquest is a consequence of both 'bottom-up' and 'top-down' processes of socio-economic and political change and continuity. The contemporary figure of the *mayora* or *cocinera tradicional* captures this ambivalent expression of a traditional Mexican cuisine which is at once inextricably tied to ancestral ingredients and techniques and powerfully conditioned by today's heritage industries and their accompanying institutional codifications of identity. There are two aspects in the forging of traditional Mexican cuisines worth highlighting for our purposes, in the process also underlining how—far from merely representing an innocent blend of cultures—modern Mexican foodways are also the product of deeply unequal social relations and uneven spatio-temporal development.

The first of these involves the peculiar forms settler colonialism took in Hispanic America, and Mexico in particular, and how this shaped subsequent food cultures. Spanish conquistadors initially aimed to reproduce in American lands the idealised peninsular diet of the noble *hidalguía*: plentiful red meat, wheat bread, olive oil, and wine (Pilcher 1996). In addition to adhering to Galenic principles of humoral diets, they sought to uphold a social hierarchy through food by, for instance, distinguishing the consumption of wheat bread and wine (integral to the Catholic Eucharist) from the corn tortilla or tamal eaten by the heathen natives (Long 2003; Earle 2014). However, the fact that these Spanish patriarchs in the main arrived to the New World unaccompanied by women led them to employ cooks and servants from indigenous or other non-European backgrounds. This in turn occasioned the gradual incorporation of autochthonous and other extra-European ingredients and techniques into Mexican creole cookery, including the signature *mole poblano*. For their part, Amerindian populations eventually took to Old World meats, citruses, and vegetables—as well as distilled spirits—thereby inventing quintessentially Mexican items such as meat-filled tacos and mezcales of global projection today. From the sixteenth century onwards, the arrival on Mexican shores of peoples from across Asia, Africa, the Pacific, and Eastern Europe added to the eclectic mix of produce, cooking styles, and dishes which, in urban contexts in particular (see, for instance, Cope 1994), have yielded all manner of culinary syncretisms, ethnic inflections, and regional specialities including African influences in dishes like cassava pasties (*empanadas de yuca*), mashed plantain (*machuca de plátano*), or pumpkin soup (*sopa de calabaza*) associated to Veracruz; the Chinese-Mexican hybrid cooking of Baja California and Sonora; or the famous *tacos al pastor*—a twentieth-century innovation imported, in cooking method at least, from the Middle East.

The settler-colonial framework thus sits uneasily with the evolution of traditional—or perhaps more accurately 'popular'—Mexican foodways since colonial efforts at codifying and policing racial and ethnic segregation became increasingly complicated over the centuries (Mörner 1967; Israel 1975). The demotic admixture of peoples and the creole cultures it engendered were further entrenched through the appropriation of *mestizaje* by Republican nation-building, which crystallised after the 1910 Revolution, and the gradual emergence thereafter of a domestic market linking different regions of the national economy (Knight 1990).

As Rebecca Earle (2007) has illustrated for the Mexican case, pre-colonial symbols, historical figures, and heroic narratives were incorporated into certain expressions of creole patriotism as early as the seventeenth century, thereby initiating a process of 'self-indigenisation' where those born in Mexican lands could claim an ancestral continuity with indigenous civilisations and in distinction to Spanish settler-colonial domination, regardless of ethnicity or 'casta'. In the aftermath of the Mexican Revolution, diverse strands of 'race thinking' tried to conjure up distinctive features of the Republic's core identity in the shape of a patronising *Indigenismo*, a separatist *Indianismo*, or perhaps most famously, in José Vasconcelos' superior, eugenic 'Cosmic Race', blending ascribed characteristics of all the country's various ethnic groups (Miller 2004; Knight 1990). Paradoxically, one achievement of the Revolution was to replace the one indigenous, agave-based alcoholic beverage that had thrived under the colonial period, *pulque*, with European beer as the dominant plebeian drink, through a combination of land expropriation of the 'old regime' *pulque* landowners and temperance campaigns that targeted the autochthonous agave brew as especially pernicious (Katz and Lazos 2017).

This does not of course mean that such hybridity and integration was everywhere smooth and evenly distributed. Plainly social stratification along class, gender, geography, and ethnicity found constant expression through, for example, the metropolitan elite's adoption of French bourgeois cuisine as a marker of distinction. (President Porfirio Díaz's 1910 banquet on occasion of the Republic's centenary is perhaps the most famous example of this—the nine-course menu contained not a single Mexican dish and was accompanied only by European wines.) It is, however, to insist that from the colonial period of New Spain, popular Mexican cuisine comprised an infinitely eclectic and hybrid offer. For urban Mexicans in particular, everyday patterns of food and drink consumption were (and continue to be) guided principally by a constellation of factors: purchasing power; whether eaten at home or at work; in a private establishment or off a street stall; in company or alone; among mixed or same-sex groups; during the day or at night; in celebration or in need. 'A Porfirian *lagartijo* (dandy)', Pilcher evocatively writes of *belle époque* Mexico, 'might eat Creole mole for *almuerzo* (brunch) at home in the morning, French *bifstek* during the *comida* (dinner) at a restaurant in the afternoon, and indigenous tacos for *cena* (supper) while cruising the streets late at night … the social boundaries around food consumption were artificial, fluid, and readily transgressed, whether by elites slumming among the lower classes or through the mingling of different plebeian groups' (Pilcher 2012: 81).

It is with this backdrop, where any gastronomic divide between settler and native, colonial and indigenous cuisine is so thoroughly compromised, that *mayoras* have to navigate their communitarian identity as traditional cooks. In an (inter)national gastronomic marketplace that seems to prize innovation and originality, *cocineras tradicionales* are interpellated via 'Ethnicity, Inc.' as purveyors of authenticity (Comaroff and Comaroff 2009). Here, etymological associations between author (*autora*), authority, and authenticity become especially relevant since the CCGM and the *mayoras* themselves place so much value upon indigenous knowledge, techniques, instruments, and practices when defining a traditional cook: Nelly, Reyna, Martha, Eloy and the *Mujeres de Humo* in this regard embody the ancestral transmission of communitarian culture. Yet this demand on their authority is simultaneously challenged by the official accreditation requirements of the CCGM and the consumer sovereignty of heritage industry clients. The fact that UNESCO's listing refers to specific practices and techniques, as well as a crops and instruments, gives further subjectivity to the type of heritage at play: that preserved and transmitted by named traditional cooks and their collaborators—overwhelmingly women, who bear the responsibility for the intergenerational reproduction of communal identity. The evolving, creative quality of the individual cooks safeguarding traditional Mexican cuisine thereby also complicates any static, essentialised understanding of indigenous gastronomy as a cultural form frozen in time. As reflected in the narratives above, traditional cooks represent personal, authorial qualities as much as they might customary communal practices, therefore making it harder to officially regulate changes, adaptations, and innovations in recipes and repertoires.

All of this leaves us with inconclusive answers to the question of how traditional Mexican cuisine, and the women who embody it, relate to settler-colonial food cultures. Plainly, as reflected in both the UNESCO submission and the activities of the CCGM, one response is that there exists in Mexico today a lasting continuity with pre-Hispanic foodways in the shape of traditional cooks and the recipes, rituals, ingredients, and culinary methods they reproduce and safeguard. Indigeneity, community, and rurality seem to be prerequisites for cooking traditional Mexican cuisine. To that extent, the latter represents not only the living memory of native cultures in general, and their gastronomy in particular, but also an expression of resilience and resistance in the face of 500 years of overseas settlement in Mexico.

Our own interpretation of the problematic of settler-colonial food cultures in contemporary Mexico on the other hand emphasises its conflicted hybridity and indeterminacy. Despite the CCGM's efforts at codifying a 'Michoacán paradigm', traditional Mexican cuisine today finds a multitude of complex and often inchoate expressions. The reflections offered by the *mayoras* we interviewed suggest a much more contingent relationship to the benchmarks of traditional Mexican cuisine, where different regional, ethnic, class, and—in the case of Eloy—gender identities are negotiated and adjusted to the expectations of both market forces and institutional pressures. This performative dimension of the role *mayoras* or *cocineras tradicionales* play in safeguarding and promoting traditional Mexican gastronomy does not seek to diminish their authority and authenticity (in the sense discussed earlier). But it does situate these claims in a wider socio-economic and political context where the structural legacies of settler colonialism relating to land tenure, collective rights, or racial hierarchies are elided, leaving a conception of indigeneity and cultural identity that tends towards atomisation and essentialisation. That is, the figure of the *mayora* as represented in contemporary Mexican cuisine becomes a retrospective construction mainly geared towards today's heritage tourism and 'culinary diplomacy'.

Bibliography

Bessière, Jacinthe, and Laurence Tibère, eds. 2011. *Food Heritage / Patrimoines Alimentaires, Revue Internationale Anthropology of Food*. http://aof.revues.org/index6759.html.

Brulotte, Ronda L., and Michael Di Giovine, eds. 2014. *Edible Identities: Food as Cultural Heritage*. Ashgate.

Comaroff, Jean, and John Comaroff. 2009. *Ethnicity Inc*. Chicago: University of Chicago Press.

Cope, Douglas R. 1994. *The Limits of Racial Domination: Plebeian Society in Colonial Mexico City, 1660–1720*. Madison, WI: University of Wisconsin Press.

Córdoba, Nelly. 2018. Una abogada con corazon de cocinera. TedxAvdelos Rios. August. Último acceso: 10 de January de 2021. https://www.ted.com/talks/nelly_cordova_morillo_una_abogada_con_corazon_de_cocinera.

de Humo, Mujeres. 2018. *Mujeres de Humo: Recetario de la cocina Espiritual Totonaca*. Mexico City: Secretaria de Cultura and Centro de las Artes Indigenas.

de Sureiman, Charles-Edouard. 2017. *Cuando la Alimentacion se Hace Patrimonio: Rutas Gastronomicas*, Globalizacion y Desarollo Local. Travauz et Recherches dans les Ameriques du Centre, 165–181.

Earle, Rebecca. 2007. *The Return of the Native: Indians and Myth-Making in Spanish America, 1810–1930*. Durham, NC: Duke University Press.

———. 2014. Diet, Travel and Colonialism in the Early Modern World. In *Global Goods and the Spanish Empire 1492–1824. Circulation, Resistance and Diversity*, ed. Bethany de Aram and Bartolome Yun-Casalilla, 137–147. Basingstoke: Palgrave Macmillan.

Gabaccia, Donna R. 2000. *We Are What We Eat: Ethnic Food and the Making of Americans*. Cambridge, MA: Harvard University Press.

Hryciuk, Renata E. 2019. La Alquimista de los Sabores: Gastronomic Heritage, Gender, and the Tourist Imaginary in Mexico. *International Latin American Studies Review* 24: 3–28.

Israel, Jonathan I. 1975. *Race, Class, and Politics in Colonial Mexico 1610–1670*. Oxford: Oxford University Press.

Katz, Esther, and Elena Lazos. 2017. The Rediscovery of Native 'Super-foods' in Mexico. In *Eating Traditional Food: Politics, Identity and Practices*, ed. Brigitte Sébastia, 20–47. New York: Routledge.

Knight, Alan. 1990. Racism, Revolution, and *Indigenismo*: Mexico 1910–1940. In *The Idea of Race in Latin America, 1870–1940*, ed. Richard Graham. Austin, TX: University of Texas Press.

Long, Janet. 2003. *Conquista y Comida* (Consecuencias del Encuentro de dos Mundos). Editado por Janet Long. Mexico City: UNAM.

Matta, Raúl. 2019. Mexico's Ethnic Culinary heritage and Cocineras Tradicionales (Traditional Female Cooks). *Food and Foodways* 27 (3): 211–231.

Miller, Marilyn Grace. 2004. *Rise and Fall of the Cosmic Race: The Cult of Mestizaje in Latin America*. University of Texas Press.

Mörner, Magnus. 1967. *Race Mixture in the History of Latin America*. New York: Little Brown.

O'Neill, Molly. 1993. Top-Notch Tables; Teotitlan del Valle, Mexico. *New York Times*, 17 January: p. 16.

Pilcher, Jeffrey M. 1996. Tamales or Timbales: Cuisine and the Formation of Mexican National Identity 1821–1911. *The Americas*: 193–216.

———. 2012. *Planet Taco: A Global History of Mexican Food*. Oxford: Oxford University Press.

UNESCO. 2010. Dossier de Candidature N° 00400 pour L'Inscription Sur La Liste Representative Du Patrimoine Culturel Immateriel En 2010. Nairobi, November.

Wilk, Richard. 2006. *Home Cooking in the Global Village: Caribbean Food from Buccaneers to Ecotourists*. London: Bloomsbury.

Yankelevich, Pablo. 2012. Mexico for the Mexicans: Immigration, National Sovereignty and the Promotion of Mestizaje. *The Americas* 68 (3): 405–436.

Zurita, Ricardo Muñoz. 2010. *Diccionario Enciclopédico de la Gastronomía Mexicana*. Mexico City: Editorial Clio.

Part II

From Erasure to Decolonisation

6

Unsettling the History of Macadamia Nuts in Northern New South Wales

Adele Wessell

Introduction

The history of macadamia nuts unsettles colonial tropes about native foods in Australia. It is the only native Australian crop that has been developed and traded internationally as a commercial food product, although Hawai'ian cultivars provide most of the world's production. This chapter takes a place-based approach to the macadamia to reveal the intersection between the unique history of the region, the environment and the culture of the Northern Rivers where the tree is endemic. While this history reveals changing attitudes to Indigenous flora and a local appreciation for the nut that disputes a lack of interest in native foods, there are continuities between contemporary food culture and colonial

I acknowledge the Traditional Custodians of the land on which I am writing and writing about, the Widjabul Wia-bal people of the Bundjalung Nation, who cared for and protected the rainforest on which macadamia grew.

A. Wessell (✉)
Southern Cross University, East Lismore, NSW, Australia
e-mail: adele.wessell@scu.edu.au

R. Ranta et al. (eds.), *'Going Native?'*, Food and Identity in a Globalising World,
https://doi.org/10.1007/978-3-030-96268-5_6

attitudes. Food has been central to the colonial project in Australia and the motivation for settling the Northern Rivers which dispossessed the Widjabul Wia-bal people. The food regime that dominated the area was directed to the production and transport of produce for the imperial market and only when demand was already established did macadamia production become commercialised. The negation of Aboriginal plant knowledge and food sovereignty from the outset hindered sharing knowledge to contribute to the sustainability of macadamia, now threatened with extinction. A reconsideration of macadamias through the lens of relationships and land is to move some way towards unsettling the history.

Unsettling History

The history of macadamia nuts both unsettles and challenges colonial tropes about native foods in Australia. It is the only native Australian crop that has ever been developed and traded internationally as a commercial food product. The tree nut crop is native to the lowland subtropical rainforest extending from south-eastern Queensland to north-east New South Wales. A local history of the macadamia reveals changing attitudes to Indigenous flora and yet continuities between contemporary food culture and colonial attitudes. The macadamia nut was eaten as a "bush nut" and experiments in commercial production began as early as the nineteenth century, disrupting conventional representations of colonial distaste for native foods as simplistic and misleading. And yet, until 1997, Hawai'i was the largest producer of the nuts in the world since the 1920s when extensive plantations as a commercial crop were established. Ironically, three species Indigenous to Australia are now on the International Union for Conservation of Nature (IUCN) Red List as threatened with extinction in the wild as they come under significant environmental pressure (Forster et al. 2020a, b, c). Management of impacts on the environment, such as chemical pesticides, soil erosion and soil health, is a challenge in the region and to community support for the industry. The establishment of the industry signals at once changes in how the land is used and Indigenous foods valued and, conversely, a continuation of colonial relations and land use patterns.

I come to this research as a white settler complicit in the ongoing dispossession of Bundjalung people and a resident of Dundarimba/Lismore. The colonisation of north-eastern New South Wales was a 'small history' in the 500-year drama of dispossession, settlement and transformation of global reach and consequence. As Vin D'Cruz reminds us, small histories and local cultures can be erased from national histories because they can stand in dissonant relation to those homogenising accounts. The forces that have shaped food history in settler colonies have influenced both the regimes that have governed people's lives as well as the environment and simultaneous attention to structural events and changes and the 'small histories', microprocesses and historical subjects who lived with and through them, bring these to life. As Christopher Mayes argues, food has been central to the settler-colonial project, 'as a necessary means of survival, but also as an avenue through which the land was possessed and a culture cultivated' (2018: 2). The history of macadamias on the North Coast provides a provocative case study in food and settler colonialism and this place-based approach is also an attempt to reconsider it in specific ways informed by relationships and land. Food production was the motivation for settling the North Coast of New South Wales and used to rationalise the dispossession of Bundjalung people without a treaty by erasing the history of Aboriginal land use in falsely declaring the land 'unoccupied'. The spread of pastoralism and agriculture also devastated Aboriginal economies. As a consequence, knowledge about native foods and their production and how to reduce the impact of consumption was also disregarded.

Henry Rous was commissioned to find fertile lands suitable for settlement in 1828 and explored the river system, naming the Richmond after Charles, the fifth Duke of Richmond. He found the area to have 'suffered as much from drought as the southern districts' and described seeing many Aboriginal people and huts, 'upwards of 30 feet in length and 6 feet in height' (Rous 1828: 355, 345). In the same year Allan Cunningham found the mouth of the river by land as he went about naming various parts of the country with his companions Logan and Fraser, to complement their friends and in memory of European places. The process of naming reinscribed the landscape with their own familiar narrative, but as Tony Birch argues, 'To name spaces is to "name histories" and also to create them' (1996: 177). Cunningham used his explorations to collect botanical specimens, proclaiming, 'I can blend discovery with botanical research tolerably well' (cited in Perry 1966). During his journeys, Cunningham also sowed various seeds into Australian soil which he had brought with him from England, Brazil and the Cape of Good Hope. 'I always carry into the interior a small bagful of peach-stones', he explained, although his journals suggest various fruit stones and seeds, 'and whenever I find a piece of good soil in the wilderness I cause it to be dug up and drop in a few in the hope of providing a meal for some famished European…or some hungry blackfellow' (cited in Lee 1925). On this trip in 1828 with Logan and Fraser, Cunningham is understood to also have taken a macadamia specimen, variously called Kindal Kindal, Boombera, Jindilli, Bauple, Gyndl, Goojabarigh, depending on the Country, having no idea that it was a food or that Aboriginal people had their own nourishment. In the Northern Rivers macadamia grew in subtropical rainforest with *Araucaria cunninghamii*, named for Cunningham, also known as *Goorumbal* or Hoop Pine, a totem of the Widjabul Wiabal people of the Bundjalung Nation.

Macadamia was named by the Victorian government botanist, Ferdinand von Mueller, in 1857 as a dedication to Dr John MacAdam, the honorary secretary at the time (and later president) of the Philosophical Institute of Victoria (Hardner et al. 2009: 10). MacAdam was the secretary of the committee that despatched Burke and Wills to Central Australia, and like Cunningham, von Mueller was a great supporter of botanical exploration and collecting. MacAdam was umpire of the first game of Australian Rules Football (AFL) in 1858 (Blainey 2010: 17). Elements of Australian Rules Football originated from a traditional game, *Marngrook*, played with a possum skin ball in various parts of Victoria under different names. The origins of the Australian game still provoke intense debate among historians that insist on denying any shared history or relations between Aboriginal and settler cultures (Hocking and Reidy 2016). The first botanical specimens named were of *Macadamia ternifolia*, although Mueller's material included what was subsequently classified as *M. integrifolia*. *Macadamia tetraphylla* was classified as a separate species in 1954 (Hardner et al. 2009: 11–12) and mainly occurs in Northern New South Wales. *M. integrifolia* is now listed as vulnerable and *M.*

tetraphylla identified as endangered in the IUCN Red List of Threatened Species, despite, and in part because of, their commercial success.

The relationship between botany and colonialism in Australia is so clear that Kamay, the place of James Cook's first landing, was named Botany Bay and planned later as the site for the penal colony. Blending 'discovery with botanical research' aligned with seeking knowledge about the economic potential of plants in the imperial project, not restricted to Australia (Baber 2016). Cunningham worked at Kew Gardens and most of his specimens were sent there, although if he sent any macadamia there it has not been studied. The first domesticated macadamia is claimed to be a tree in the Brisbane City Botanical Gardens planted in 1858 by Walter Hill, then superintendent of the Gardens (who also worked at Kew from 1843 to 1851), now registered with the National Trust (Hardner 2016: 1412). As Babar explains, networks of botanic gardens that were coordinated by the central node or Kew Gardens constituted the complex integrated links that were essential for the global networks that enabled the mobility of plants, people, power, profit and patronage that were the indispensable ingredients of the emergent science of botany (668).

At the local level, colonial botanists were also complicit in the control of Aboriginal people's lives. Marcia Langton has traced the relationships between colonial botanists and natural scientists with the work of Aboriginal Protectors. Underpinning this were dual concerns: introducing and establishing economic plants, and identifying edible native plants and methods to prepare these to reduce toxicity (Langton 2016). Walter Hill was engaged in both these endeavours at the Botanical Gardens. The combination of observation and description and the reduction of plants to a scientific classification system that did not include local knowledge meant that recognition of macadamia as providing food was delayed for some time. This discovery is popularly awarded to a boy, who suffered no ill consequences from consuming the nut (McConachie 2012: 5). Aboriginal people enjoyed macadamia nuts for thousands of years. Stone technology and plant processing tools used to crack nuts and grind kernels have been dated to c3500 years in Queensland archaeological sites (Cosgrove 1996: 905).

It took around thirty years after Cunningham's specimen was collected before anyone was recorded as eating the macadamia nut. Settlers rarely had strong enough relationships with Aboriginal people, relationships built on the trust and respect required to observe, ask questions, mimic and try new things. For country cooks like Mina Rawson native produce was a necessary part of the diet, but she acknowledged in *Antipodean Cookery Book and Kitchen Companion* (1895) that Aboriginal people had shared with her their knowledge of food and medicines. Rawson encouraged people to experiment and 'try everything', advising (1895: 54): 'Whatever the blacks eat the whites may safely try'. As grazing and agriculture expanded into Aboriginal land, exotic food production became more stable reducing the need for local supplies but also encroaching on the Aboriginal plant food economy.

The factors that restrained settlers and explorers from knowing Australian plants as food and enjoying them are multifarious. Believing native foods to be 'wild-harvested' and uncultivated enabled colonists to see the land as unoccupied and, as Callicott reminds us, available to 'make over into a landscape like the one they left behind' (2000: 24). It was a convenient ignorance that served the colonial project and settler culture. Cultural determinism and food racism have been cited by John Newton as underlying the rejection of native foods (2016: 89–92) and that, conversely, eating can be an act of culinary reconciliation. A similar argument often draws on multiculturalism equating the expansion of Australians' palates with respect. Neither satisfactorily explains local appreciation of native foods, documented in nineteenth-century cookbooks and newspapers, which could not be seen as a reflection of more enlightened attitudes. The consumption of native foods in the Northern Rivers was actually quite widespread. The food regime that dominated the Northern Rivers, however, was directed to the production and transport of produce for the imperial market emphasising the exchange value of food for which taste and demand were already established. Successful Australian and state governments did little to promote or create a market for new products.

The history of macadamias reflects another deeply held Australian insecurity, a case of cultural cringe. As a writer in the *Sunday Times* pointed out in 1905, 'Possibly when America begin to export them to

Australia we will begin to appreciate them' (1905: 1). In the late nineteenth century macadamia was introduced into Hawai'i, and the first large-scale commercial development occurred there. As in Australia, the trees were initially planted as ornamentals and for home gardens and as windbreaks, but recognition of the palatability of the kernels motivated the establishment of orchards from the 1920s. The development of reliable grafting methods and scientific research by the University of Hawai'i and Hawaiian Agricultural Experimental Station culminated in the release of cultivars and germplasm developed in Hawai'i which is now the major source of the current world production of macadamia (Hardner 2016). Hundreds of *M. tetraphylla* were planted in the 1920s, but *M. integrifolia* became the preferred species for commercial production, also known as 'rough' and 'smooth' shell nuts respectively. Efficient cracking machines provided a breakthrough.

Colonial Cultivation

Before and during the dominance of Hawai'i in commercial trade of the nut, there were local voices raised in support of an Australian industry, unable to rouse government support comparable to Hawai'i. The first cultivated orchard of macadamia trees, consisting of the local *M. tetraphylla*, was planted by Charles Staff in 1882 at Rous Mill, twelve kilometres from Lismore in the North Coast of New South Wales. When the nuts were still plentiful in Northern New South Wales and Queensland, there were advocates for the quality of the local nut writing in the *Northern Star*, *The Richmond River Herald*, *The Sydney Mail*, *the Daily Mercury* in Mackay, *the Maryborough Chronicle*, the *Central Queensland Herald* and the *Eastern Recorder*. Where macadamias grew, the locals did have some sense of their value. As the *Australian Town and Country Journal* declared in 1886, 'There should, at least, be one Australian nut tree grown in every large garden and moderate sized orchard, in suitable places…from Jervis Bay to the Tweed, both for the sake of its nuts and for its fine ornamental appearance' (1899: 29). From a local perspective, the rejection, neglect and the rise and popularity of the macadamia looks very different from the national record of the nut.

Beyond the colonial economic imperative of producing food for export, the macadamia was appreciated at a very local level. Walter Hill began growing macadamia for distribution in 1867 and wrote letters to the editor publicising the 'handsome tree' in the hope of drawing the attention of cultivators (1867: 3). By this time *The Queenslander* reported, 'The nut is now well known to the timber-getters, to the natives and others, and quantities are being daily gathered and eaten, thus proving its wholesome qualities. Of the quality of the fruit we can speak in the highest terms' (1867: 11). Local trees at this stage were still found in abundance. In 1870 the Queensland government forbid the cutting and removal of timber from macadamia and bunya pine trees on vacant Crown Land and pastoral leases (Department of Public Lands 1870: 1). Land clearing since European development from the nineteenth century, however, significantly impacted the diversity of macadamia. They have limited capacity for dispersal and to expand their habitat, existing in isolated pockets or at the fringe of rainforests (McConachie 2012: 2). The alienation of Crown Land in New South Wales proceeded rapidly during the 1870s and 1880s, accelerating the dispossession of Aboriginal people and limiting their land and economy. While macadamia habitat in Northern New South Wales in the 'Big Scrub' was largely selected but only slightly cleared in the mid-1880s, it was almost entirely removed by about 1920 (Stubbs 2001: 296). Stray trees, however, were sometimes left in the clearings, because people already realised how good the nuts were. A gap between commercial imperatives and taste can be observed. It is also the case that the colonies in Australia were established for the purpose of supplying the empire and the task of creating a market was beyond settlers used to considering anything Indigenous to be inferior.

In small orchards and home gardens locals could still covert the macadamia nut. Native plant foods were less familiar (and had less status) than local animal products in the diets of the British. Australia has an exceptionally high level of endemism among plants, most often not found elsewhere. Local fruits, such as Davidson's Plum and Riberries were prepared by settlers in ways that were familiar and palatable to them in jams and cordials. But as Ian McConachie explains, 'Macadamias were regarded as a very special food and held in high esteem as our own true nut. Its delicate distinctive flavour and crunchy texture distinguished it from other tree nuts...', but 'unfortunately for Australia we tended to talk about these wonderful nuts but did little to address what was needed to grow and process them as a commercial crop' (2012: 10–11). There were repeated calls to cultivate the crop commercially, dating from the late nineteenth century and recorded in the *Agricultural Gazette* as well as newspapers. 'The story is a very sad one', the curator of the Technological Museum declared. If the advice offered by this museum had been followed in 1930, when an officer was sent to the North Coast for the express purpose of stimulating interest in the cultivation of this native species, Australia today would have been in a position to export large quantities of the nut, besides meeting domestic needs (*Cairns Post* 1941: 8).

At least part of the problem might have been the confusion about the nuts created by the name, variously called bush nuts, bopple or bauple nuts, Queensland nuts, Mullumbimby nuts. Macadamia, on the other hand, was the American trade name from the beginning, having no need, obligation or understanding to acknowledge local histories. The banality of the name would not alert consumers to its origins and thus distanced the macadamia from its habitat. At the first conference of nut growers and manufacturers of nut products convened in 1932, delegates voted to use the 'Australian nut', based on national sentiment and colonial appeal, 'as the nut would be a fine advertisement for the Commonwealth' (Hewitt 1932: 4). The society proposed a levy, but small membership would not have provided sufficient funds for investment or support for the new name. Meanwhile, the global industry continued to grow with crops in Africa and Central and South America being established. Up until 1970 the total crop never exceeded 70 tonnes of nut in shell per year in Australia, while Hawai'i averaged about 250 tonnes per year (McConachie 2012: 13).

North Coast macadamia production expanded rapidly in the 1970s, a period of significant transition in the region. With the end of dairying, a gradual but radical transformation was underway. New settlers, many seeking alternative lifestyles and new ways of living generally streamed in. From the early 1970s a 'population turnaround' began along the warm eastern seaboard, remaining largely unrecognised by government planners until well into the 1980s. The movement of young adults and young retirees out of the southern cities and inland to the north coast of New South Wales and south coast of Queensland is the most recent migration into the area. During the 1970s and 1980s a substantial number of these new settlers moved to the abandoned dairying land of the hinterland valleys to form multiple occupancy communities or buy individual blocks for new farming ventures. Like many of their pioneering predecessors of the nineteenth century, the new settlers of the 1970s and 1980s who took up farmland were often ignorant of the land ecology, climatic environment and farming practices and started with little. However, they came with a different land ethic and often clashed in the early decades with the established farming communities. Dave Forrest, who arrived in 1978 recalls being determined not to repeat the mistakes of the past in terms of large-scale land clearing and chemical use and set up with the intention of organic farming, including a macadamia orchard (Forrest 2010).

The Australian Macadamia Society was formed in 1974 to represent growers, undertake research and promote the macadamia nut. Funding for research and promotion expanded in the 1980s through the levy they established, matched by the Australian government and tax minimisation schemes also encouraged industry expansion. Farms grew and the industry expanded and Australia is now the world's major producer of macadamia nuts. Australia and South Africa are the largest producers of macadamia nuts (55% of total world production), followed by Kenya, China, the USA (Hawai'i), Guatemala, Malawi, Vietnam, Colombia, New Zealand and Swaziland (CBI 2021).

While cultivated orchards of macadamia are extensive, they are almost genetically uniform. The macadamia industry is based primarily on hybrid cultivars and Hawaiian cultivars are the main contributors to world production. Genome sequencing by Catherine Nock and Craig Hardner suggests that seeds from a single tree or a couple of trees taken from Gympie in Queensland were the foundation of the Hawaiian macadamia industry which now supplies about 70% of the world's macadamia varieties (Nock et al. 2019). Fragmentation between uncultivated trees and cross-pollination from orchards poses a threat to genetic diversity. While the geographic origin of many domesticated crops can be difficult to trace, populations of macadamia were relatively undisturbed prior to the nineteenth century. As one of the most recent domesticated international crops, they are only two or three generations from this history with cultivars selected through breeding programmes in Hawai'i released less than 100 years ago.

Most of the habitat of macadamias in the North Coast has been lost, and with that, the diversity so lacking in the industry. Local remnant populations of macadamia and their rainforest habitat are subject to intense pressure from development in addition to habitat modification by weeds and fire regimes. Pesticide dependency is a major challenge in the industry and this may be in part due to the scale of operations, the monoculture they epitomise and the lack of predators that would have been present in their natural habitat. Ironically, given the native status of macadamia, impacts on the environment, particularly pesticide dependency and soil erosion, are a big challenge in the area and support for the industry is not secure.

While a growing enthusiasm for local foods proceeds alongside the successful establishment of the export industry in the nuts, attention to the relationships between the local and global in environmental and economic terms is frequently neglected. Moreover, the provenance of native ingredients is often absent. As Charlotte Craw has pointed out, notions of the local are often conflated with the national in packaged foods that identify 'Australian native ingredients' (2012a) and this can de-Indigenise food within a national culinary culture by not showing connection with place. The diversity of native foods across different locations and seasons, the origin of ingredients and their connections with Indigenous Australian cultures and places are often omitted, marginalising Indigenous knowledges and conceptions of place. Protection of Indigenous rights to intellectual and cultural property has only recently been established, but information about plant use already in the public domain provide no benefit for Aboriginal knowledge holders (Cunningham et al. 2009: 432).

The survival of the industry and the species may ultimately depend on access to trees grown in their original local habitat and the hundreds planted in gardens and backyard orchards from the late nineteenth century. Coordinated conservation of the macadamia was initiated by the Australian Macadamia Society in 1988 that instigated the formation of the Macadamia Conservation Trust and recovery plan for all four species (Healthy Land and Water 2020). While isolated trees still exist in rainforest scrubs, the 'Wild Macadamia Hunt' now also depends on the distribution of seedlings Walter Hill encouraged. Trees of around 100 years old may be the sole descendants from populations that no longer exist. Restoring some genetic diversity to macadamias is a necessary response to climate change, new diseases, pests, the survival of the species and, ultimately, the industry.

This definition of 'wild' includes reference to trees intentionally planted in the movement to promote the macadamia evidenced in the local newspapers which served Queensland and Northern New South Wales. Notions of 'wilderness' and 'wild' foods more generally though have effectively removed Aboriginal people from the landscape and agriculture. As Deborah Bird Rose explains,

> Aboriginal people in Australia and elsewhere have rightly objected to a definition of wilderness that excludes human impacts because it necessarily excludes them and all the generations of their people who lived in country and took care of it. (2012: 9)

Patricia O'Connor, Yugambeh Elder, recalls macadamia nuts being planted.

'When I was a little girl, probably seven or eight years old I was cracking Queensland nuts', eighty-eight-year-old Patricia O'Connor told NITV. 'My grandmother said "when I was a little girl I planted those nuts as I walked with my father along the Nerang river" and she said "you call them Queensland nuts, I call them Goomburra". 'She planted them when she walked with her dad, and as an adult she saw them bearing fruit' (cited in Archibald-Binge 2017).

This disputes the history of macadamia as regenerating unaided prior to colonisation. In contrast to the understanding of wild as pristine and untouched and empty of people, the term is highly problematic when talking about land inhabited for tens of thousands of years. In contrast, 'Wild people (colonisers) make wild country (degrading, failing)' (Rose 2004: 4). Cultivated macadamia orchards with all the associated problems of pesticide dependency and soil erosion could also, in this light, be considered 'wild'.

A great diversity of food plants have been identified in the subtropical rainforests where macadamias once grew (Beck and Balme 2003). Food production in these areas compromised native food sources and introduced species settled into sometimes uneasy relations in the local ecology. From the 1970s and 1980s an interest and taste for native foods of the Northern Rivers has expanded. Alongside the establishment of the macadamia industry, the cultivation and harvesting of Lemon Myrtle, Aniseed Myrtle, Davidson's Plum, Riberry, Native Mint, Wild Limes, Warrigal Greens, Dorrigo Pepper and Finger Limes has also developed commercially. Support for Indigenous people to create their own businesses, facilitate access to their own cultural food and recognise the knowledge and customs associated with growing and preparing native foods remains an issue.

One of the implications of not acknowledging Aboriginal cultivation was the lack of attention paid to knowledge about reproduction and care and how to reduce the impact of consumption. The slow development of macadamia as a commercial food crop can be partly attributed to ecological factors: the large number of native insect pests and diseases and inability to control them (McConachie 2021). For sixty years commercialising a rainforest tree into a monoculture made the industry unviable although they are endemic to the country. There was, at the same time, a lack of effort in Australia, which the Hawaiians put in, to learn how to graft the trees and manage them.

The history of macadamias provides a lens through which to consider settler colonialism. Cultural, economic, ecological and material forces have intersected to contribute to the fate of these rainforest trees in the places where they are endemic. Local foods were a necessity in early settlement, but the clearing of land for exotic species which simultaneously impacted Aboriginal plant economies threatened the survival of the trees, even as this was recognised in government limits to their removal. While loss of traditional knowledge about the use and the ecologies of macadamias would accompany this, indifference and deliberate negation of Aboriginal plant knowledge from the outset also contributed losing knowledge necessary to ensure their sustainability. While the records of botanists have provided a rich source of ethnohistorical material, they also documented factors that led to the destruction of Aboriginal plant economies which relate to macadamias. In the words of Marcia Langton, 'Their records represent…a window on the competition of the most educated colonists for the resources that would support ever-expanding herds of cattle and food for the colonists and the English market' (2016: 221). Scientific endeavours included the documentation, naming and sometimes dispersal of the nut tree, even as the focus in the case, for example, of Walter Hill was to disseminate information most useful to graziers and farmers. The emphasis placed on the introduction of species established for world trade transformed the environment into a 'productive' colony. As Gascoigne and Curthoys explain (2002):

> This ecological revolution began with the First Fleet which, thanks to [botanist Joseph] Banks, included on board a microcosm of the agriculture of Britain and its empire. Scientific selection and techniques of botanical transportation were put to use to refashion the new world of Australia into a botanical image closer to that of the old world of Europe…With ecological implications that are still being worked out, as much of the land as possible was turned over to the cultivation of introduced crops and the support of non-indigenous animals which came in increasing numbers by the late 1790s. The macadamia industry has now been recentred in the Northern Rivers, but the biggest single income earner in agriculture is still beef cattle, a potent symbol of English identity.

Final Thoughts

Contemporary interest in native foods represents colonial continuity, growing on stolen ground with little involvement of Aboriginal people. Only 1% of the industry is thought to involve Indigenous people (Mitchell and Becker 2019). Economic imperatives continue to drive the relationship to land. Local politics has been traditionally dominated by a party that had its roots in 'countrymindedness' but is now firmly aligned with mining interests that frequently diverge from the farmers they once represented. Demand for change is growing from settler consumers opting for more sustainable and authentic products, but this does not necessarily extend to changes in power relations or land ownership. Local appreciation for macadamias in the past still coincided with the violence of colonialism and lack of inclusion, consultation, acknowledgement of Indigenous people and land persists. The contemporary interest in native foods is often symbolised as a departure from the past, where a lack of appreciation is conflated with colonial attitudes (Craw 2012b). Revealing the taste for macadamias in the past that coexisted with colonialism also suggests it can exist as a structure in the present. More often than not the Aboriginal history presented in relation to macadamia is firmly embedded in the past erasing the authority of Aboriginal people living on Bundjalung Country today. The stories of macadamia are told through individual settlers. Embracing native foods can be used in the construction of a national cuisine, an imagined connection and belonging to place still predicated on erasing Indigenous food sovereignty and appropriation. It is what Grey and Newman would define as culinary colonialism: 'the extension of Settler jurisdiction over, and exploitation of, Indigenous gastronomy' (2018).

The challenge of any collection of settler narratives is that the local and small voices can be subsumed in a grand narrative as if the power of the coloniser is uniform across the expanse of the colonies they have claimed possession of. The potential they have, on the other hand, is to elevate the impact of shared experience of that power, the indifference to the local that comes from having status inferred from elsewhere, where significance was held to transpire. Macadamia was never universally disregarded or overlooked. The obstacles to growing the local industry are both distinct to local history and shared across the nation. In the Northern Rivers the local ecology and features of the macadamia were particular challenges. Shared across Australia, however, indifference to local foods and knowledge about them and the absence of government support and investment in research is of ongoing concern. 'Australia' functions as a nation as a product of colonial history and convenience. On the ground it is a diverse and complex place, where different languages once described the distinctive conditions and histories of what matters. In a global economy, terms like macadamia may communicate across boundaries and languages to sell produce bereft of its history or the cost of that to Indigenous people, to the species and the environment.

Bibliography

Archibald-Binge, Ella. 2017. Elders Take Centre Stage at Buckingham Palace. *NITV News*, March 4. Accessed April 1, 2021. https://www.sbs.com.au/nitv/nitv-news/article/2017/03/14/elders-take-centre-stage-buckingham-palace.

Baber, Zaheer. 2016. The Plants of Empire: Botanic Gardens, Colonial Power and Botanical Knowledge. *Journal of Contemporary Asia* 46 (4): 659–679. https://doi.org/10.1080/00472336.2016.1185796.

Beck, Wendy, and Jane Balme. 2003. Dry Rainforests: A Productive Habitat for Australian Hunter-gathers. *Australian Aboriginal Studies* 2: 4–20.

Birch, T. 1996. A Land so Inviting and Still Without Inhabitants: Erasing Koori culture from (post-) colonial landscapes. In *Text, Theory, Space: Land, Literature and History in South Africa and Australia*, ed. K. Darian-Smith, E. Gunner, and S. Nuttall, 173–188. London: Routledge.

Blainey, Geoffrey. 2010. *A Game of Our Own: The Origins of Australian Football.* Melbourne: Black Inc.

Bush, Farm & Garden. 1867. *The Queenslander*, April 6, 11.

Callicott, J. Baird. 2000. Contemporary Criticisms of the Wilderness Idea. *USDA Forest Service Proceedings.* 1: 24–31.

CBI. 2021. The European Market Potential for Macadamia Nuts. Accessed March 2021. https://www.cbi.eu/market-information/processed-fruit-vegetables-edible-nuts/macadamia-nuts/market-potential

Cosgrove, Richard. 1996. Origin and Development of Australian Aboriginal Tropical Rainforest Culture: A reconsideration. *Antiquity* 70: 900–912.

Craw, C. 2012a. Tasting Territory: Imagining Place in Australian Native Food Packaging. *Locale: Pacific Journal of Regional Food Studies* 2: 1–25.

———. 2012b. Gustatory Redemption? Colonial Appetites, Historical Tales and the Contemporary Consumption of Australian Native Foods. *International Journal of Critical Indigenous Studies* 5 (2): 13–24.

Cunningham, A.B., S.T. Garnett, and J. Gorman. 2009. Policy Lessons from Practice: Australian Bush Products for Commercial Markets. *GeoJournal* 74: 429–440.

Department of Public Lands. 1870. Government Advertisement, *Brisbane Courier*, May 26, 1.

Forrest, Dave. 2010. Interview with Hazel Ferguson, Landed Histories Project. Southern Cross University.

Forster, P., Griffith, S., Ford, A., and Benwell, A. 2020a. *Macadamia integrifolia. The IUCN Red List of Threatened Species*. Accessed March 29, 2021. https://doi.org/10.2305/IUCN.UK.20203.RLTS.T113180064A113310165.en.

Forster, P., Ford, A., Griffith, S., and Benwell, A. 2020b. *Macadamia Tetraphylla. The IUCN Red List of Threatened Species*. Accessed March 29, 2021. https://doi.org/10.2305/IUCN.UK.20203.RLTS.T113180222A113310175.en.

Forster, P., Ford, A., Griffith, S., and Benwell, A. 2020c. *Macadamia Ternifolia. The IUCN Red List of Threatened Species* Accessed 29 March 2021. https://doi.org/10.2305/IUCN.UK.20202.RLTS.T113180135A113310170.en.

Gascoigne, John, and Patricia Curthoys. 2002. *The Enlightenment and the Origins of European Australia*. Cambridge: Cambridge University Press.

Grey, Sam, and Lenore Newman. September 2018. Beyond Culinary Colonialism: Indigenous Food Sovereignty, Liberal Multiculturalism and the Control of Gastronomic Capital. *Agriculture and Human Values* 35: 1–14.

Hardner, C. 2016. Macadamia Domestication in Hawai'i. *Genetic Resources and Crop Evolution* 63: 1411–1430. Accessed September 10, 2020. https://doi.org/10.1007/s10722-015-0328-1.

Hardner, Craig M, Cameron Peace, Andrew J. Lowe, Jodi Neal, and Phillip Pisanu. 2009. Genetic Resources and Domestication of Macadamia. *Horticultural Reviews* 35: 1–125. Accessed March 29, 2021. https://media.johnwiley.com.au/product_data/excerpt/28/04703864/0470386428.pdf

Healthy Land and Water. 2020. The Wild Macadamia Hunt. Accessed October 13, 2020. https://hlw.org.au/project/the-wild-macadamia-hunt/?fbclid=IwAR3MXbdw1xzGAihJCQcBhokj8QsD7zocgYRR0YDyjjucBk9fGlABAN28rgw

Hewitt, N.C. 1932. Nut Culture Association Formed Great Possibilities. *Northern Star*, July 16, 4.

Hill, Walter. 1867. New Fruit Indigenous to Queensland. *Brisbane Courier*, March 6, 3.

Hocking, Jenny, and Nell Reidy. 2016. Mangrook, Tom Wills and the Continuing Denial of Indigenous History. *Meanjin*, Winter. Accessed March 20, 2021. https://meanjin.com.au/essays/marngrook-tom-wills-and-the-continuing-denial-of-indigenous-history/

Langton, Marcia. 2016. Botanists, Aborigines and Native Plants on the Australian Frontier. In *Land and Language in Cape York Peninsula and the Gulf Country*, ed. Jean-Christophe Verstraete and Diane Hafner, 221–240. Amsterdam: John Benjamins Publishing Company.

Lee, Ida. 1925. *Early Explorers in Australia*. Project Gutenberg of Australia eBook. Accessed December 14, 2020. http://gutenberg.net.au/ebooks03/0301141h.html

Macadamia Nut Neglected Australian Species. 1941. *Cairns Post*, September 22, 8.

Mayes, C. 2018. *Unsettling Food Politics: Agriculture, Dispossession and Sovereignty in Australia*. London: Rowman & Littlefield.

McConachie, Ian. 2012. *The Macadamia Story* Lismore: The Australian Macadamia Society.

———. 2021. Personal Communication. June 28.

Mitchell, R., and J. Becker 2019. Bush Food Industry Booms, But Only 1 Per cent is Produced by Indigenous People. *ABC News*. Accessed June 30. https://www.abc.net.au/news/rural/2019-01-19/low-indigenous-representation-in-bush-food-industry/10701986#:~:text=Australia's%20bush%20food%20industry%20is,to%20an%20ongoing%20research%20project.

Newton, John. 2016. *The Oldest Foods on Earth: a history of Australian Native Foods with Recipes*. Sydney: NewSouth Publishing.

Nock, C.J., C.M. Hardner, J.D. Montenegro, A.A. Ahmad Termizi, S. Hayashi, J. Playford, D. Edwards, and J. Batley. 2019. Wild Origins of Macadamia Domestication Identified Through Intraspecific Chloroplast Genome Sequencing. *Frontiers in Plant Science* 10: 334. https://doi.org/10.3389/fpls.2019.00334.

Perry, T. M. 1966. Cunningham, Allan (1791–1839). *Australian Dictionary of Biography*, National Centre of Biography, Australian National University. published first in hardcopy 1966. Accessed March 27, 2021. https://adb.anu.edu.au/biography/cunningham-allan-1941/text2323

Rawson, Mrs. Lance. 1895. *The Antipodean Cookery Book and Kitchen Companion*. Kenthurst: Kangaroo Press.

Rose, D. 2004. *Reports from a Wild Country: ethics for decolonisation*. Sydney: UNSW Press.

———. 2012. Why I Don't Speak of Wilderness. *Earthsong Journal Perspectives in Ecology, Spirituality and Education* 2 (4): 9–11.

Rous, Henry. October 1828. A Description of the Rivers Clarence and Richmond in latitude 28 degrees. 9m. and 28 Degrees 53m. Respectively, &c. in New South Wales. From Recent Observations. *The Australian Quarterly Journal of Theology, Literature & Science* 4: 352–355.

Stubbs, Brett J. 2001. The 'Grasses' of the Big Scrub District, North-eastern New South Wales: Their Recent History, Spatial Distribution and Origins. *Australian Geographer* 32 (3): 296.

Sunday Times. 1905. September 17, 1.

The Australian Nut. 1899. *Australian Town and Country Journal*, May 6, 29.

7

Definitions of Hawaiian Food: Evidence of Settler Colonialism in Selected Cookbooks from the Hawaiian Islands (1896–2021)

Laura Kitchings

Printed Material and Settlement Colonialism in Hawai'i

This chapter uses five cookbooks to trace the evolution of settler colonialism in the Hawaiian Islands. This chapter is not an attempt to explore all of the foodways existing in the Islands, but to examine how definitions, and who controlled these definitions, of Hawaiian food evolved with the structure of settler colonialism. The first two cookbooks present initial developments in the defining of Hawaiian foodways as a result of settler colonialism. The third cookbook shows how the importation of a labour force to serve the needs of the original settler colonialists led to a redefinition of Hawaiian foodways. The final two cookbooks demonstrate attempts to decolonialise the Islands' foodways and current acknowledgement of Hawai'i as a settler-colonial state. The research relied solely on English language materials available to me as a white woman in

L. Kitchings (✉)
North Grafton, MA, USA
e-mail: kitching@bu.edu

R. Ranta et al. (eds.), *'Going Native?'*, Food and Identity in a Globalising World,
https://doi.org/10.1007/978-3-030-96268-5_7

Massachusetts, living on the traditional territory of the Nipmuc Nation in 2020 and 2021.

There were multiple newspapers in the Hawaiian language, serving Native Hawaiian interests, in existence soon after the 1820 arrival of the missionaries from the Massachusetts-based American Board of Commissioners for Foreign Missions (ABCFM). These newspapers often served as tools for Native Hawaiian resistance to American interests, especially after the ABCFM stopped funding the descendants of the missionary families in the mid-nineteenth century. The use of Hawaiian language newspapers as a form of active resistance against American interests in the Islands led to the 1896 banning of Hawaiian language education in schools and prohibitions against using the Hawaiian language in public. The probations were not overturned until the Hawaiian Renaissance beginning in the 1970s. As the Chair of the Indigenous Studies Program at McGill University Noelani Arista discussed in a virtual lecture hosted by Rare Book School headquartered at the University of Virginia in Charlottesville, there are now a number of initiatives to work with and train Hawaiian speakers to index and catalogue historical Hawaiian language materials. More accessible Hawaiian language materials will result in more nuanced scholarship about the evolution of multiple Hawaiian foodways and the development of the settler colonialist structure in Hawai'i.

In *American Settler Colonialism*, Walter L. Hixson (2013) estimates the Native Hawaiian population to be at least 400,000 individuals when Captain Cook first visited the Islands in 1787. In 1823, the Native Hawaiian population in the Islands was about 135,000 individuals, and by 1893 (the year of the coup by American interests against the Hawaiian monarchy), the Native Hawaiian population is estimated at 40,000 individuals. While some Native Hawaiians left the Islands for employment, the decline in population was largely due to diseases brought to the Islands by Westerners.

The US Census (2021) estimates that in 2019 there were 1.43 million residents of Hawai'i, with 876,154 living on the most populated of the eight major islands, O'ahu, many in and around the capital city of Honolulu. Approximately 25% of Hawaiian residents identify as white, 37% identify as Asian, and 10% identify as Native Hawaiian or other

Pacific Islander. Approximately 24% of Hawaiian residents identify as two or more races and 18.5% were born outside of the United States. According to the Hawai'i State Department of Education (2021), this census data includes 100,000 active military personnel and their families. According to the Hawai'i Tourism Authority (2020), in 2019, 10.4 million people visited the Islands, spending US$17.75 billion and supporting 216,000 jobs

Cookbook Studies

As discussed in the works of Cookbook Scholars Jessamyn Neuhaus (2012), Megan Elias (2017), and Emily Contois (2020), cookbooks reflect the society of their creation. They can be read as both prescriptive and aspirational domestic literature. Working with cookbooks may not provide insights to how specific individuals in a community thought about daily cooking or their enjoyment of domestic labour. While notes written in individual cookbooks may provide information about their use, the text of the cookbooks defines what foodways were presented as prescriptive and/or aspirational in the featured society.

Additional care needs to be taken when working with community cookbooks. The reader cannot assume that the recipe contributors to community cookbooks have first-hand experience with a presented recipe. The cookbooks display the collective public-facing voice of the community over the voices of individual contributors. The language used in recipes often reflects the domestic experience, scarcity of certain ingredients, and economics of the community. The community cookbooks offer insights depicting heritage, lifestyle, and values of the cookbook creators that otherwise may be undocumented, not necessarily documentation of the foods eaten within the individual households. It displays how the cookbook contributors want to be presented to the public as part of a defined community (Bower 1997).

Hawaiian Cookbook (Ladies Society of the Central Union Church: 1896)

This first cookbook examined for this chapter is the 1896 fourth edition of the *Hawaiian Cookbook* produced by the Ladies Society of Central Union Church and published by the Hawaiian Gazette Company Print, the same company that published the pro-American business interests English language newspaper *The Hawaiian Gazette*, a weekly and then a semi-weekly published between 1865 and 1918 (Library of Congress 2021). The Introduction to this fourth edition of the cookbook states the superior cooking abilities of the "American housewives of Honolulu" demonstrating that they likely saw themselves retaining their American identity while living most of their lives in the Islands.

As Joy Schulz explains in *Hawaiian by Birth* (2017), the ABCFM declared the Islands Christianised in the mid-nineteenth century and ended their financial support of the missionary families and their descendants. However, by this time the families had settled in the Islands and considered the Islands their multi-generational permanent home. The descendants of the missionary families were not considered American citizens and many had been raised with elite Native Hawaiian children, however did not see the Native Hawaiian occupants of the Islands as their equals. When visiting the United States Mainland, the descendants were not embraced for their Americanness. This may have led to them rallying behind their own identity as Americans among themselves. While some of the contributors descended from elite Native Hawaiian families, most of the contributors descended from intermarriages between missionary families (Schulz 2017)

Most of the recipe contributors are identified at Mrs [Husband's First Initial] [Husband's Last Name]. While the identification of the recipe contributors by their husbands' names can be read as simply maintaining American traditions, given the stress on heteronormativity and prescribed gender roles that was as part of interactions between American female missionaries prescribing heteronormative gender roles to Native Hawaiian women starting from the arrival of the missionaries (Kauanui 2018).

The cookbook also maintains the Western tradition of daily cooking as a married woman's responsibility.

The majority of the recipes in the cookbook represent what Food Historian Rachael Laudan (2015) terms a middle-class Anglo cuisine and are similar to recipes that could be found in cookbooks throughout the British and American Empires at the time. These recipes include potato pancakes, raised biscuits, and brown bread. Most of the ingredients used in these recipes required ingredients regularly imported from the Mainland.

There is a presented menu for "A Hawaiian Feast", in a separated part of the cookbook. The section is credited to a "Mrs. Brickwood and Her Daughters", a family group connected to missionary descendants, Native Hawaiian and Chinese merchant families. The menu introduces Native Hawaiian food preparations and food names. It includes instructions for how the menu should be served, including being eating on the ground. In this cookbook it is unclear if it is meant as cultural appreciation or appropriation. However, it places the menu as a foodway unrelated to appropriate daily food preparations in Honolulu and differentiated from the normalised recipes using imported food.

The 1920 paperback sixth edition of the cookbook presents the same section but adds a section on local fish from the Women's Committee of The Territorial Food Commission and Federal Food Administration pamphlet dated June 1918. Here, use of the Native Hawaiian names of the fish has been divorced from the Native Hawaiian culture and the fish are meant to be used in recipes in order to limit meat imports to the Island during World War I. Instead of being celebrated, the local fish have been militarised, demonstrating the increased militarisation of the Islands. The white-coded church group is now not only using the fish but stating how they should be used for American military efforts, claiming authority over the Native Hawaiian identification of the fish.

Throughout the cookbook, the women present their versions of recipes primarily prepared using imported ingredients as the proper food to be served in Honolulu. While a few named Native Hawaiian ingredients are incorporated into specific recipes, it largely presents American-affiliated women as owning the foods prepared in the Islands and the experts on these foods, noting Native Hawaiian knowledge in select circumstances.

The cookbook presents foodways replicated from the contributors' largely New England familial origins as defined Hawaiian foodways.

How We Serve Hawaiian Canned Pineapple (Hawaiian Pineapple Packers' Association [HPPA]: 1914)

The 1914 edition of the Hawaiian Pineapple Packers' Association (HPPA) corporate cookbook, *How We Serve Hawaiian Canned Pineapple*, presents a different definition of Hawaiian food developed by settler colonialists that eventually serves the tourist industry. The non-native pineapple became the food most associated by outsiders with the Islands. This small book, according to its catalogue record, of twenty pages with illustrations measures thirteen by fifteen centimetres. Examining a digital surrogate of the book, the front cover has an illustration of a pineapple field without workers, a large pineapple next to an open can, and a plate of pineapple slices. It conveys that the pineapple goes directly from the field to its sliced presentation in the can with minimal human intervention. The back cover features featureless men with a variety of hats; some of the hats suggest a connection between the workers and East Asian nations. The Introduction places the pineapple as a Hawaiian plant and the pineapple fields as a place of beauty. The Introduction continues to introduce the product as not expensive and instructs the reader on the proper way to open the can.

The corporate cookbook minimises the colonial history of canned pineapple, placing it as an ingredient found in, not brought to, the Islands. While eventual head of the HPPA, Harvard University graduate James D. Dole, was not the first to grow pineapples in Hawai'i, he used his connections to investors on the Mainland to successfully produce and develop a market for the canned Caribbean 'Smooth Cayenne' variety of pineapple, determined to be the best fit for the Islands by horticulturalist John Kidwell.

Dole, arrived in Hawai'i, soon after the 1898 American annexation of the Islands with plans to use the connections of his relative, American business advocate Sanford B. Dole, to obtain work with one of the "Big Five" Corporations in Hawai'i. The five networked sugar plantation corporations, Castle & Cooke, Alexander & Baldwin, American Factors, C. Brewer & Co., and Theo. H. Davis dominated Hawai'i's economy for decades following annexation. These corporations took advantage of the reorganisation of land ownership in Hawai'i to a Western-style system in 1848 to develop sugar plantations, which eventually consumed the Islands.

James D. Dole purchased land at Wahiawa, O'ahu, using loans from his Boston-based father, and new land acts making Island lands available for white settlers. He and his Harvard classmate George Damon Dutton experimented on the land, planting crops that would not compete in the market with the successful crops produced by neighbouring Chinese farmers.

By 1901 they found success in growing and producing canned pineapples for the Mainland United States market. While there had been pineapple canning previously in the Hawaiian Islands, Dole's Hawaiian Pineapple Company (HPC) was able to develop the cannery on a larger scale due to investors on the United States mainland and was able to use Sanford B. Dole's political and financial connections to employ Chinese labourers for the company.

There is no easily accessible documentation on pineapple exports from Hawai'i for 1901, but by 1903 Dole's Hawaiian Pineapple Company had produced 1893 cases of canned pineapple. Between 1907 and 1909, pineapple sales dropped significantly, and the Hawaiian Pineapple Growers' Association (HPGA) was founded to create a demand for canned pineapple. This organisation had some success in creating collective marketing strategies. HPGA was disbanded in December of 1912 and was immediately replaced by the Hawaiian Pineapple Packers' Association (HPPA). The new group funded pineapple agricultural research into breeding a pineapple that was resistant to mealybug wilt, a common problem among the Hawaiian pineapple companies. The group also realised the need for new marketing strategies for their product (Okihiro 2009).

The HPPA soon partnered with the expanding *American Ladies' Magazine* market and members of the developing Home Economics movement. This marketing in American middle-class publications took advantage of concerns over adulteration of foods produced in the Mainland United States, stressing that the canned pineapple process was sterilised and sanitary, specifying the minimal human contact was used in the production of the product. This cookbook, part of this marketing strategy, presents pineapple recipes from popular white American cookbook authors, some taken directly from published cookbooks of the time. The recipes would have likely been familiar to the audience of the *Ladies' Magazine*, such as salads and fritters. In many recipes the pineapple simply replaces the more familiar apple. The presentation of the pineapple in a can followed the new national scientific standards of nutrition and hygiene in Mainland United States foodways, and the idea housework could be made much easier through the use of industrialised food following proper instruction, both championed by evolving Home Economics field. Canned pineapple became marketed as a useful and inexpensive ingredient while it also exoticised everyday cooking if housewives followed some easy instructions from the cookbook.

This 1914 corporate cookbook places pineapple as a natural part of the Islands, not part of the developing industrialisation of the Islands to serve American business interests. This aligned with how the Hawaiian Islands began to be marketed as an empty pristine landscape, with modern conveniences, as a playscape for white Mainland United States residents (Miller-Davenport 2019)

As the Hawaiian tourism industry flourished, with tourists first arriving by steamship and later by air, the Hawaiian tourist industry developed a performative form of Hawaiian history for tourist consumption. These events that primarily benefited American business interests did not include explanations of pre-contact food rules defining gendered food interactions, physical changes to the Islands caused by plantations, or the changing populations of the Islands as contract labourers began to come to the Islands to work on the plantations. These events featured the introduced pineapple as central to Hawaiian foodways.

The distinct shape of pineapple allowed for easy identification in visual culture and was soon featured as a visual element on advertising materials for the Islands including steamship menus on ships bringing the tourist to the Islands. Eating or drinking pineapple products on the Mainland became a substitute for visiting the Islands, an inexpensive way to relive or imagine the tourist experience in Hawai'i (Klein 2020).

In 1932, Castle & Cooke, who held a minority share in James D. Dole's company, forced through a plan of reorganising the company, resulting in a new company owned by Castle & Cooke. Dole separated from the new company and briefly served as Chief of the Food Division of President Franklin Delano Roosevelt's Agricultural Adjustment Administration (AAA). After leaving the governmental post, he joined a gold dredging venture in California and eventually started the food development company James Dole & Associates.

A legacy of James D. Dole was to normalise a definition of the pineapple as the quintessential Hawaiian food—a Hawaiian food that could easily be procured on the United States Mainland for a low cost. The marketing served to place Native Hawaiians in service to the tourism market and detach them from the agricultural processes in the Islands. Therefore, the agricultural use of the Islands was presented in service to Mainland residents, not the Native Hawaiian population.

Favorite Island Cookery: Book V 1889–1989 (Honpa Hongwanji Hawaii Betsuin: 1989)

The third cookbook examined in this chapter is a spiral-bound 2005 reprint of the 1989 *Favorite Island Cookery* published by the Honpa Hongwanji Hawaii Betsuin. It represents the fifth cookbook published by Honpa Hongwanji Mission of Hawai'i and celebrates the 100-year anniversary of the organisation in the Islands. The cookbook contains 228 pages of recipes, maps of Japan and Hawai'i, and a definition and discussion of the importance of *Sojin Cookery* (Japanese Buddhist foodways) to this community. The recipes are not credited to specific individuals, but some are credited to specific locations, or as "Mother's Favorite". One hundred and twenty-eight of the recipes are credited to specific locations in Japan and Okinawa, forty-one recipes are credited as "Mother's Favorite", thirty-eight are not credited with a location, and two hundred and twenty-six recipes are credited to "Hawaii".

Pineapple and sugar plantations, mainly run by the Big Five, significantly changed the economy and geography of the Islands. They also changed the population demographics of the Islands. The aggressive industrialised agricultural development of the Islands serving American-identified business interests discussed in the last section required a significant labour force.

A 1975 article in the *Hawaiian Journal of History* (Char and Char) traces the South China origins of the first group of Chinese contract labourers who arrived in the Islands in 1852. While the 1886 report of the Hawaiian Bureau of Immigration lists the ship *Thetis* as departing from Hong Kong, the article confirms Amoy (Xiamen), Fukien (Fujian Province), as the departure point for this first group of labourers bound for the Islands.

Arnold Hiura's *Kau Kau: Cuisine and Culture in the Hawaiian Islands* (2009) traces the immigration and foodways of the Chinese labourers and some of the many labour groups that followed once they arrived in Hawai'i. His work estimates that between 1894 and 1900, 57,000 Japanese labourers arrived in Hawai'i (the first Japanese labourers arriving in 1868). Between 1900 and 1920, 2095 Puerto Rican labourers migrated to the Islands. Okinawan labourers first arrived in 1900 and their immigration statistics are included with the Japanese immigration figures, and between 1903 and 1905, 7843 Korean labourers arrived to work on plantations. In 1906, the first Filipino labourers arrive, and by 1916 there are 18,144 Filipino labourers in the Islands.

Hiura's work describes the generalised layouts of the plantation towns that were often within miles of each other by the 1950s and 1960s. The towns consisted of labourers' houses clustered around a mill with the waste products sent into the ocean. The towns generally had a public school, a post office, garage and machine shop, barbershop, gas station, baseball fields, gyms, and churches. Some towns had a dispensary for medical needs, a bakery, and a theatre. These towns had at least one store, run by the corporation who owned the plantation. Eventually former labourers started businesses around the plantations that carried staples including canned goods, bread, and canned meat. Most of the plantations allowed space for small gardens, many labourers kept their own chickens, and there were some small dairies and chicken farms that sold milk and eggs.

Ronald Takaki's foundational work on daily plantation life, *Pau Hana* (1984), describes how the plantation class worked to control all aspects of the labourers' lives and discourage worker unification. Most plantations consisted of a tiered housing pattern, related on distance from the sewage system to re-enforce a culturally defined hierarchy. While the housing conditions varied from plantation to plantation, they were generally crowded and unsanitary, and medical services varied by plantations were often hampered by communication and cultural barriers to care.

The plantation groups employed various groups of labourers, not only due to immigration regulations, but also to take advantage of pre-existing histories of distrust among the labour populations to avoid labour unification and strikes. After the United States annexed Hawai'i, the immigration laws prohibiting Chinese immigration now applied to the new territory. The plantation owners worried that as the Japanese labourers became the majority on the plantations, they would successfully organise against the working conditions. The networked sugar plantation owners decided to employ Korean labourers, assuming that the previous history between the two East Asian nations would discourage unification of the two groups of labourers on the plantations. Takaki also discusses other cultural groups that came as labourers, most significantly the Portuguese labourers who arrived in the Islands planning to settle in the Islands, unlike the other groups of labourers who planned on returning to their home countries after they fulfilled their contracts.

Plantation owners allowed for spaces and occasionally gave financial support for religious spaces, programmes, and leaders such as Buddhist temples, programmes, and priests. Plantation managers viewed the temples and programmes (and other religious spaces and programming) as a stabilising influence on the workers. It is also possible that they saw religion as another way to keep the various groups of labourers separated to avoid unified labour action.

Kelli Y. Nakamura's article in *The Hawaiian Journal of History* (2018) describes the limited documentation available on the important role played by Japanese women in the establishment of Buddhism in Hawai'i. The article explores the various public-facing and domestic roles filled by these women and how these roles encouraged the women of the community to exert agency while pursuing personal and communal goals. The article discusses the possible loneliness of many Japanese women in the plantation culture and how women's groups based on a shared religious identity allowed for a communal identity. The article touches on the many responsibilities of the women belonging to these groups during World War II, as many of the men were either active in the US military or sent to Japanese internment camps on the United States Mainland. The article also states that these groups often re-enforced stereotypical gender and family norms.

The Honpa Hongwanji Hawaii Betsuin cookbook follows the tradition of Japanese immigrant women to the United State Mainland creating cookbooks in the first half of the twentieth century that served to maintain community while managing cultural adaption to American products. The cookbooks also served to re-enforce traditional gender roles, focusing on the importance of married women to provide regular, culturally appropriate meals to their families (Matsumoto 2013).

Several of the recipes in the Community Cookbook credited to Hawai'i include American brand name ingredients such as Ritz crackers, or ingredients generally associated with the foodways of other groups brought over as part of the plantation labour force. A few recipes include Japanese ingredients such as yomogi leaves cooked in the microwave, following the evolution of kitchen technology found in many American homes. The recipes throughout the book display a communal voice of home cooks creating and defining a unique Hawaiian-Japanese-American foodway.

One surprise presented when working with the recipes in this community cookbook was the large number of recipes that included milk as an ingredient. As Rachel Laudan explains in her foundational work on foodways in Hawai'i, *Food of Paradise* (1996), American Home Economics instructors were a fixture in the Hawaiian public school system throughout the twentieth century and followed nutritional advice prevalent on the United States Mainland. This meant that Home Economics teachers encouraged their students to eat a more standard Mainland diet, including a significant consumption of milk, likely indigestible to many of the descendants of the plantation workers. This assumption that the amount of milk suggested relates to Mainland dietary guidelines becomes more probable when examining the 1995 version of *Favorite Island Cookery* which discusses drinking skim milk due to fat concerns in regular milk and explains the usefulness of nutritional labels used on the Mainland United States of the time.

While many of the plantation labourers did not plan to stay in the Islands, their descendants became a new group of settler colonialists on the Islands. Their foodways have become more associated with Hawai'i than the foodways of Native Hawaiians. In her essay "Settlers of Color and 'Immigrant' Hegemony: 'Locals' in Hawai'i", Native Hawaiian Activist Haunani-Kay Trask (2008) addresses that while those descended from plantation workers have faced racism and suffering in the Islands, Hawaiian identity should be controlled by Native Hawaiians. The essay clearly delineates between identities of "local" and "Native Hawaiian", with locals defined as descendants of non-white workers on the plantations. Previously, those descended from plantation workers have claimed to include Native Hawaiians in the definition of local as a way to separate their (those descended from plantation workers) history on the Islands from that of white residents of the Islands. As the next two sections will discuss, this discussion over the definition of local extends to definitions of Hawaiian foodways.

Taste Our Love for the Land: Recipes and Sustainability Stories from the Chefs of the Hawai'i Food Wine Festival (Denise Hayashi Yamaguchi: 2015)

The fourth cookbook discusses recent changes in the restaurant culture of the Islands. The cookbook celebrates the Hawai'i Food and Wine Festival in a hard-bound coffee-table-style book with professional photography. This Festival was an outgrowth of the Hawaiian Regional Cuisine movement discussed later in this section.

The narratives of the foodways that developed on the plantations have often been flattened to labourers sharing rice-based foodways which developed into a unified working-class Hawaiian cuisine. However, as shown in the last section, many of the plantation labourers and their descendants worked to root their new Hawaiian-based foodways in their traditional foodways.

The realities of plantation life did require the evolution of the labourers' traditional foodways. The works of Arnold Hiura, drawing on his own personal experiences in the plantation town of Papa'ikou on the Big Island of Hawai'i, detail how the realities of plantation life melded with the previous food traditions of the labourers to form new Hawaiian-based foodways. Hiura also traces the closures of most of the plantations between 1969 and 2010 as the corporations moved their operations elsewhere and the new labourers found by former plantation workers.

The labourers and their families learned to utilise the canned meat sold around the plantations, deal with the reality of shipping strikes that threatened the labourers' reliance on imported food and the food rations of World War II, and manage and share the harvests of their personal gardens. In many cases new similar food traditions were adapted into the foodways of the various cultures of the labourers. For example, many dishes evolved to incorporate the canned meat product Spam (Hiura 2013).

As many plantation labourers stayed in the Islands after leaving the plantations, the 1950s and 1960s saw the rise of dinner-style restaurants and lunch wagons that continued the narrative of universal food sharing on plantations to sell what is now defined as "local" food. As local-style lunch wagons and restaurants expanded, foods unique to these establishments flourished such as the Loco Moco (a bowl of hamburger meat, rice, gravy, and often a fried egg) and the plate lunch (a protein, two scoops of rice, and macaroni salad) that claimed to relate to multiple foodways that evolved among plantation workers. Currently videos available on the fast-casual popular Hawaiian-based restaurant chain Zippy's website (FCH Enterprise, Inc. 2021) discuss the origin of "Saimin" (a noodle and broth soup) and how it relates to soups present in Chinese, Japanese, Okinawan, and Filipino foodways. This working-class restaurant cuisine became what is now referred to as local Hawaiian food and can be found at a variety of restaurants throughout the Islands. It is differentiated from the foodways found in the homes of many descended from plantation workers defined as locals.

This discussion of local food becoming the dominant food identity of the Islands is both similar and different from how pineapple came to define the foodways of the Islands. Pineapple as the quintessential Hawaiian food served the tourist industry and the plantation class. Local food as the quintessential Hawaiian food serves to create a unified identity for those locals living on the Islands but arguably erases the foodways of the Native Hawaiian population as Hawaiian food.

Until the early 1990s these foods were the foodways of the working class in Hawai'i descended from plantation workers, while up-scale restaurant hotels served imported food in French-styled menus. In her book on food and empire Rachel Laudan (2015) discusses how French high cuisine spread throughout a variety of empires, marketing it as the pinnacle of food in the nineteenth and twentieth centuries. Teaching and explaining French cuisines became a large industry, and professional chefs in upper-class restaurants were expected to have formally trained in France. This was true in the post-contact Hawaiian Islands as restaurant culture developed among the first settler-colonialists.

In his chapter "Hawai'i Regional Cuisine" published in *Eating Asian American,* Samuel Hideo Yamashita (2013) explains that starting in 1991, chefs associated with the Islands began meeting to consider how to lessen the reliance of Hawaiian restaurants on imported food and lessen the French influences on fine dining in Hawai'i. Modelled after other regional cuisine movements of the 1990s, the chefs partnered with local farmers, fishermen, and other food purveyors bringing awareness to the decimation of the Islands brought by the now closing sugar and pineapple plantations. The movement sought to elevate local food, and dishes inspired by the working-class restaurants appeared on high-end restaurant and hotel menus. However, as Yamashita notes, the movement did not centralise the foods of the Native populations of the Islands, but of the groups in Hawai'i descended from former plantation workers and continued the dominant definition of Hawaiian foodways as a unified cuisine developed on the plantations. The essay does not feature a discussion of pineapple, as the movement likely did not seek to alienate tourists by depriving them of the Hawaiian tourist staple of pineapple discussed earlier in this chapter.

The cookbook features stories of chefs loosely associated with Hawai'i. It also includes recipes featuring products farmed in Hawai'i and examples of projects to reverse the harms of industrial agriculture on the islands such as the restoration of Native Hawaiian fishponds. There is also a focus on the need for chefs worldwide to source more sustainable seafood. The cookbook is a celebration of the actions of the chefs but does not discuss the harms continuing on the islands or how hotels (that include the restaurants of many of these chefs) continue to harm the Islands through the tourist industry.

In many ways the cookbook can be seen as an attempt to decolonise Hawaiian foodways by de-centralising the restaurant foodways of the original white settler-colonialists and attempting to restore localities on the Island harmed by the industrialised agriculture caused by plantations. This also includes lessening Hawaiian dependence on foods imported to the Islands. However, many of the stories in the cookbook feature model projects from outside the Islands to serve as inspiration for Hawaiian-based projects. While the cookbook serves as an inspiration for more international restaurant foodways featuring sustainable and restorative ecological practices, it does not centre on Native Hawaiian-led projects. The book makes an attempt to discuss decolonisation of foodways in Hawai'i without a full interrogation of who is a settler colonist in the Islands and what a true decolonisation of Hawaiian foodways would involve. Instead, it celebrates a significant change in largely high-end restaurants in Hawai'i to serve more sustainable ingredients in dishes related to the foodways of the various groups and shared restaurant culture of those considered "local". Again, not fully interrogating who counts as a local in Hawai'i and who controls this definition.

Cook Real Hawai'i (Sheldon Simeon and Garrett Snyder: 2021)

The final cookbook examined for this chapter was released as I was finishing a draft of this chapter in March 2021. It is an approximately 300-page hardbound book with professionally shot photographs of landscapes of Hawai'i and the various prepared recipes. Sheldon Simeon, a repeated participant in the American televised cooking competition, *Top Chef*, presents recipes of foods associated with the Hawaiian Islands. On the first page of the Introduction, he explains that although he lives in Hawai'i, he is not Hawaiian but descended from Filipino plantation workers. He clearly spells out the difference between being from Hawai'i (local), as descendants of labourers, and being Native Hawaiian. This may mark a new way that those creating public-facing works about foods in Hawai'i acknowledge that their definition of Hawaiian food is not a universal definition and that multiple foodways exist in the Islands. In the Introduction he introduces the reader to phrases in Hawai'i Creole English (also known as Hawaiian Pidgin), a common language in the Islands that originated on the plantations, and begins to unpack the various definitions of Hawaiian food that influenced his recipes. He incorporates the legacy of foods that were used to define the Islands for tourists, such as pineapple, but also showcases foodways of various families that draw on their own unique heritage to develop foodways that reflect the realities of living on the Island with their family histories.

The book discusses the variety of industries present in the post-contact Islands including labourers involved with cattle ranching on the Big Island of Hawai'i and the foodways of these Paniolos and their descendants who worked with the cattle. This discussion includes the foodways of these groups and how their foodways may have affected other foodways present in the Islands.

The book presents the reality of multiple foodways, some shared by multiple groups, he has experienced in the Islands. He also explains that food terms may have different definitions on different Hawaiian Islands, alerting the reader to additional divisions in Hawaiian foodways. The cookbook also discusses how the foodways of new immigrant and migrant groups to the Islands may allow for the development of new Hawaiian foodways and evolutions in the foodways of these groups as they adapt to the realities of life in the Islands.

The book serves to expand the public-facing definitions of Hawaiian food while acknowledging the history of settler colonialism by multiple groups in the Islands. Following the *Hawai'i Regional Cuisine* movement, the book chooses to focus on non-imported foods, which may reflect changes in the shared foodways of the Islands.

Conclusion

Food Studies is an interdisciplinary field that includes issues of food politics, memory, identity, nutritional needs, and financial realities. Cookbooks can reflect many of the thoughts related to these issues prevalent in the society of their creation. This chapter focuses on how cookbooks reflect how changes, including changes based on developments of dominant industries and population changes, drive changes in defined foodways of a settler-colonial state. Each of the definitions of Hawaiian food in these cookbooks only serve a small group of stakeholders related to life on the Islands. All of these stakeholders benefit in some way from being affiliated to a settler-colonial state. All of the cookbooks discussed in this chapter de-centralise the role of Native Hawaiians in defining the public-facing definitions of Hawaiian food. This needs to change if true efforts to discuss what a possible decolonisation of foodways related to the Islands might require of all participants in the Island's foodways. As climate change brings increasing changes to the Islands not considered by the original settler-colonialists to the Islands, the foodways of Hawai'i will need to continuously evolve. Hopefully, public definitions of Hawaiian foodways will begin to focus on serving all residents of the Islands while including Native Hawaiian voices.

Bibliography

Bower, Anne L. 1997. Bound Together: Recipes, Lives, Stories, and Readings. In Recipes for Reading: Comments, Cookbooks, Stories, Histories, ed. Anne L. Bower, 1–14. Amherst: University of Massachusetts Press.

Char, Tin-Yuke, and Wai Jane Char. 1975. First Chinese Contract Laborers in Hawaii, 1852. *Hawaiian Journal of History* 9: 128–134.

Contois, Emily J.H. 2020. *Diners, Dudes, and Diets: How Gender and Power Collide in Food Media and Culture*. Chapel Hill: The University of North Carolina Press.

Elias, Megan J. 2017. Food on the Page: Cookbooks and American Culture. Philadelphia: University of Pennsylvania Press.

FCH Enterprise, Inc. 2021. Saimin Says. Accessed June 1, 2021. https://www.zippys.com/saimin-says/.

Hawai'i State Department of Education. 2021. Military Families. Accessed May 1, 2021. https://www.hawaiipublicschools.org/ParentsAndStudents/MilitaryFamilies/Pages/Home.aspx.

Hawai'i Tourism Authority. 2020. Hawai'i Visitor Statistics Released for 2019. Accessed May 1, 2021. https://www.hawaiitourismauthority.org/news/news-releases/2020/hawai-i-visitor-statistics-released-for-2019.

Hawaiian Pineapple Packers' Association. 1914. *How We Serve Hawaiian Canned Pineapple*. Honolulu: Hawaiian Pineapple Packers' Association.

Hiura, Arnold. 2009. *Kau Kau: Cuisine & Culture in the Hawaiian Islands*. Honolulu: Watermark Publishing.

———. 2013. *From Kau Kau to Cuisine: An Island Cookbook, Then and Now*. Honolulu: Watermark Publishing.

Hixson, Walter L. 2013. *American Settler Colonialism: A History*. Basingstoke: Palgrave Macmillan.

Honpa Hongwanji Hawaii Betsuin. 1989. *Favorite Island Cookery: Book V 1889–1989*. Honolulu: Honpa Hongwanji Hawaii Betsuin.

———. 1995. *Favorite Island Cookery: Book VI*. Honolulu: Honpa Hongwanji Hawaii Betsuin.

Kauanui, J. Kēhaulani. 2018. *Paradoxes of Hawaiian Sovereignty: Land, Sex, and the Colonial Politics of State Nationalism*. Durham: Duke University Press.

Klein, Shana. 2020. *The Fruits of Empire: Art, Food, and the Politics of Race in the Age of American Expansion*. Oakland: University of California Press.

Ladies Society of Central Union Church. 1896. *Hawaiian Cook Book*. 4th ed. Honolulu: Hawaii Gazette Company.

Laudan, Rachel. 1996. *The Food of Paradise: Exploring Hawaii's Culinary Heritage*. Honolulu: University of Hawai'i Press.

———. 2015. *Cuisine and Empire: Cooking in World History*. Los Angeles: University of California Press.

Library of Congress. 2021. About the Hawaiian Gazette. Accessed June 3, 2021. https://chroniclingamerica.loc.gov/lccn/sn83025121/.

Matsumoto, Valerie J. 2013. Apple Pie and Makizushi: Japanese American Women Sustaining Family and Community. In *Eating Asian America: A Food Studies Reader edited by Robert Ji-Song-Ju*, ed. Martin F. Manallansan IV and Anita Mannur, 255–273. New York and London: New York University Press.

Miller-Davenport, Sarah. 2019. *Gateway State: Hawai'i and the Cultural Transformation of American Empire*. Princeton: Princeton University Press.

Nakamura, Kelli Y. 2018. The Power Behind the Scenes [Enno Shitano Chikaramochi]: The Activism of Buddhist Women in Hawai'i. *Hawaiian Journal of History* 52 (November): 89–115.

———. 2012. *Manly Meals and Mom's Home Cooking: Cookbooks and Gender in Modern America*. Baltimore: Johns Hopkins University Press.

Okihiro, Gary Y. 2009. *Pineapple Culture: A History of the Tropical and Temperate Zones*. Berkeley: University of California Press.

Schulz, Joy. 2017. *Hawaiian by Birth: Missionary Children, Bicultural Identity, and U.S. Colonialism in the Pacific*. Lincoln: University of Nebraska Press.

Simeon, Sheldon, and Garrett Snyder. 2021. *Cook Real Hawai'i*. New York: Clarkson Potter/Publishers.

Takaki, Ronald T. 1984. *Pau Hana: Plantation Life and Labor in Hawai'i 1835–1920*. Honolulu: University of Hawai'i Press.

Trask, Haunani-Kay. 2008. Settlers of Color and 'Immigrant' Hegemony: 'Locals' in Hawai'i. In *Asian Settler Colonialism: From Local Governance to the Habits of Everyday Life in Hawai'i*, ed. Candace Fujikane and Jonathan Y. Okamura, 170–194. Honolulu: University of Hawai'i Press.

United States Census Bureau. 2021. Quick Facts: Hawai'i. Accessed March 23, 2021. https://www.census.gov/quickfacts/HI#qf-headnote-a.

Woman's Society of Central Union Church. 1920. *Hawaiian Cook Book*. 6th ed. Honolulu: Honolulu Star-Bulletin, Limited.

Yamaguchi, Denise Hayashi. 2015. *Taste our Love for the Land: Recipes and Sustainability Stories from the Chefs of the Hawai'i Food Wine Festival*. Honolulu: Watermark Publishing.

Yamashita, Samuel Hideo. 2013. The Significance of Hawai'i Regional Cuisine in Postcolonial Hawai'i. In *Eating Asian America: A Food Studies Reader*, ed. Robert Ji-Song-Ju, Martin F. Manallansan IV, and Anita Mannur, 98–122. New York and London: New York University Press.

8

Decolonising Israeli Food? Between Culinary Appropriation and Recognition in Israel/Palestine

Ronald Ranta and Daniel Monterescu

Our food never received full recognition and even among the culinary elite, I never saw a clear acknowledgement that 'this is yours, it's obvious'.
—Palestinian chef Habib Daoud

Introduction: Contesting Rootedness

Over the past three decades, Israeli food has emerged as a global culinary phenomenon with Israeli cookbooks, restaurants, and chefs winning numerous awards in Israel and across the world (Ranta and Prieto-Piastro

R. Ranta
Department of Politics and International Relations,
Kingston University, Kingston upon Thames, UK
e-mail: Ronald.Ranta@kingston.ac.uk

D. Monterescu (✉)
Department of Sociology and Social Anthropology,
Central European University, Vienna, Austria
e-mail: monterescud@ceu.edu

R. Ranta et al. (eds.), *'Going Native?'*, Food and Identity in a Globalising World,
https://doi.org/10.1007/978-3-030-96268-5_8

2019). This emergence has naturally led to studies examining Israeli food as a cultural field and analysing its history and evolution (see, e.g., Avieli 2017; Grosglik 2017, 2021; Hirsch 2011; Prieto-Piastro 2017; Raviv 2015; Zaban 2016). One of the interesting aspects of Israeli food is that it is one of a very few cultural fields in which Jewish-Israelis and Arab and Palestinians openly and publicly engage with each other, though from different positions of power, as food scholars, journalists, chefs, kitchen workers, business proprietors, food producers, and consumers. As a result, the emergence of Israeli food as a global phenomenon has brought with it culinary questions regarding heritage, authenticity, and terroir, but also wider political and sociological questions regarding cultural appropriation and decolonisation.

These questions have grown louder in recent years as food dishes and ingredients, which are and have been part of Arab regional and local Palestinian food culture and heritage, have increasingly been entering Israeli food culture. The list is extensive and includes dishes such as *maqlouba* (rice, vegetables and chicken served upside down), *shishbarak* (meat dumplings cooked in yoghurt), and *knafeh* (sweet cheese pastry) and ingredients such as *khubeizah* (mallow), *'akub* (gundelia), and *laban jamid* (dry yoghurt). While this is not a new phenomenon,[1] the pace and scale of incorporation have increased, and whereas in the past the adoption and appropriation of Arab and Palestinian food were downplayed or conveniently 'forgotten', increasingly they are acknowledged as hallmarks of a new culinary identity (Ranta 2016). This acknowledgement of Arab and Palestinian food, culinary influences, chefs, and restaurants in the field of Israeli food is one of three important culinary trends that are interlinked and that we, and others (Avieli 2017; Grosglik 2017, 2021), have observed and identified. The other two include an emerging focus on terroir in Israel/Palestine, with an emphasis on locality, artisanal production, and organic produce (Grosglik 2017, 2021; Grosglik et al. 2021; Monterescu and Handel 2019), and an inclusive and flexible approach to what constitutes Israeli food, in terms of ingredients, techniques, influences, and dishes.

[1] Several food scholars (Hirsch 2011; Mendel and Ranta 2014; Raviv 2015) have argued that what is considered Israeli food has been based, to an extent, on the adoption and appropriation of Arab and Palestinian food.

Based on fieldwork conducted by us together and separately over the past decade, interviews with leading Israeli and Palestinian chefs and food scholars and writers, and a series of culinary panels we helped organise,[2] we use this chapter to re-examine the relationship between Israeli and Arab and Palestinian food cultures and politics.[3] The central question we ask is whether the field of Israeli food is undergoing a paradigmatic shift with regard to the role and place of Arab and Palestinian people and food cultures and what this might mean for Israeli settler colonial identity, society, and attempts at decolonisation more broadly. In many ways what we are seeking to understand is, what does a process of culinary decolonisation look like? Is such a process happening? And if so where, how and by whom?

In answering this question, due to the rapidly changing nature of what is perceived as or termed 'Israeli food', we do not attempt to provide irrefutable empirical evidence in support of one position or another. Instead, we engage with the main culinary narratives and discourses that we encountered, regarding the role and place of Arab and Palestinian food culture in the cultural field of Israeli food. Our focus is specifically on the actors who, by function of their position, are directly involved in the creation and maintenance of these discourse and narratives, such as chefs, restaurant owners, and food scholars and writers. This focus has a number of drawbacks, which we fully accept; these relate, among others, to the exclusive nature of many of the restaurants involved that cater to a select clientele.

Based on the research we have done, we have identified three particular discourses, which have some overlap, adopted by leading figures in the field of Israeli food with regard to Arab and Palestinian food:

[2] In the summer of 2020, alongside others, we helped organise a series of panels in English, Hebrew, and Arabic on food and politics entitled 'Turning the Tables: Palestinian and Israeli Food on the Cutting Board', under the auspices of the Van Leer Institute in Jerusalem. See: https://youtube.com/playlist?list=PLHgwT2MoGd80qZunxvfnEHz-m2cPrY3Yp.

[3] It is important to emphasise that we have no intention of essentialising these terms. Particularly with regard to Palestinian food culture, we fully acknowledge that it has been and is influenced by a wide range of factors. According to Palestinian Chef Fadi Kattan, 'Palestinian cuisine cannot be separated from the cuisine of the region. Palestinian cuisine is the result of this small piece of land occupied by different civilisations'. In terms of Arab food culture, we fully accept that there are several different and overlapping terms used, such as Levantine food, Shami food, and Middle Eastern food.

cosmopolitan appropriation, Arab-Jewish revival, and reflexive decolonisation.[4] In the sections below, we first discuss the evolution and history of Israeli food culture and its relationship with settler colonialism and with Arab and Palestinian food. We then move on to explore and engage with the three discourses mentioned above. In the conclusion we return to our main question and consider how the developments and discourses we have encountered in Israeli food culture relate to decolonisation.

Setting the Table: Settler Colonialism and Self-Indigenisation

A recent online competition by Keren Kayemeth LeIsrael-Jewish National Fund (KKL-JNF),[5] for which around 150,000 votes were cast, selected the olive as Israel's national tree (KKL-JNF 2021). On the surface this is neither surprising nor controversial. The olive tree has been a feature of the land of Israel/Palestine for thousands of years; it features prominently in the Bible, in both the Old and New testaments, and Israel's official national emblem features olive branches. Nevertheless, national symbols, whether official or otherwise, are not merely mundane images of the nation. They are descriptive and prescriptive representations that speak to prevailing discourses, power relations, and processes of inclusion and exclusion. The interpretation of these symbols is not arbitrary, but one that 'is learned and ingrained' in the collective imagination (Elgenius 2011; see also Ranta and Mendel 2016).

What does the choice of the olive as Israel's national tree represent? To start with it, the choice of a fruit tree, one that is part of the history and culture of the land and its people, speaks to the importance of food, land, and agriculture in Israel and to culinary representations of the state and

---

[4] It is important to note here that these three discourses were among those who do acknowledge and mention Arab and Palestinian food and their influence on Israeli food culture. There are many in the field who still do not.

[5] KKL-JNF was established in the early twentieth century for the purpose of facilitating Zionist settlement in Palestine through land purchases. After the creation of the state of Israel in 1948, its role was reconfigured to managing state lands, which include lands seized by the state following the 1948 war, which is referred to as the *Nakba* (catastrophe) by Palestinians.

the nation. Many of Israel's official and unofficial national symbols are food related, from the Jaffa Orange to the prickly pear (Abufarha 2008; Bardenstein 1998; Mendel and Ranta 2014). In this regard, the choice of the olive tree is also interesting and controversial because of its symbolism in the context of Jewish-Israeli and Arab-Palestinian relations. The olive tree and its main product olive oil are central features in Arab regional and local Palestinian terroir, foodways, and food culture; this also includes the historical culinary traditions of many Jews living in the region. Specifically regarding Palestine, the tree has long been a prominent feature of Palestinian communal life, farming, and economy and has emerged over the years as an important national and social symbol. The olive signifies the Palestinian attachment to the land and, in the West Bank and Gaza Strip, steadfastness (*sumud*) in opposition to Israeli occupation and land expropriation (Meneley 2011). In this context, the uprooting of olive trees, particularly in the West Bank, has become a form of collective punishment meted against Palestinians by the Israeli army and Jewish settlers (Gutkowski et al. 2013).

The choice of a tree native to Palestine, which is strongly connected to Palestinian terroir, history, and heritage, also points to some of the particularities of Israeli/Zionist settler colonialism and the importance of Zionist self-indigenisation (Veracini 2010). While there are clear similarities between Zionism and other settler colonial movements, in terms of their desire to dispossess and overcome, in one way or another, the indigenous population (see, e.g., Shafir 1989), there are also several key differences, for example, the late emergence of Zionism, in the late nineteenth century, the lack of a metropole, and the strong sense of victimhood shared by Zionist settlers, who in many cases escaped from persecution (Zreik 2016); Edward Said (1979) described Palestinians as 'the victims of the victims'. However, in this chapter we will focus on four crucial differences and explain how these manifest in Israeli food culture.

First, unlike other settler colonial enterprises that emanated from Europe, Zionism was by its very nature a rejection of the traditions and cultures in which it emerged; the term used in Hebrew is *shlilat hagalut* (negation of the exile: Raz-Krakotzkin 1993). Zionist settlers emigrated to Palestine to create not only a Jewish homeland but also a new Jewish identity and society (Almog 2000). The new Jewish identity, labelled 'muscular Judaism' by prominent early Zionists, was to be constructed in opposition to the *Ashkenazi* (European) diaspora Jew, who was seen as physically weak and mentally and culturally feeble (Mendel and Ranta 2014). In terms of food culture, this meant a rejection of traditional *Ashkenazi* food (or in the words of Claudia Roden: 'foods of exile and martyrdom' 1999). As a result, *Ashkenazi* food is largely absent from Israeli food culture and is considered 'Jewish' rather than 'Israeli'. The Zionist and later Israeli food culture that emerged was based instead on a flexible idea of what the national food should be, a feature that has remained ever since. The emerging food culture emphasised Jewish ownership of agricultural lands; the establishment of agricultural settlements; and the importance of Jewish food production and produce, which in many cases went hand-in-hand with official Zionist policies of 'Hebrew labour', namely boycotting Palestinian food produce and excluding Palestinian labour (Mendel and Ranta 2014; Shafir 1989). Additionally, the flexibility in what was termed Israeli food demonstrated an early and inherent tension between Zionist desires to emulate European non-Jewish cuisines and the growing use of local produce and the adoption and appropriation of regional and local dishes and ingredients. This tension could be seen in the fascination and aversion early settlers had towards local foods, such as olives and aubergines, and their ambivalence towards once cherished European imports (Raviv 2015).

Second, while self-indigenisation is common to all settler colonial movements, the way it manifests in Zionism is unusual. The Zionist movement did not present itself as a strictly settler colonial endeavour. Although, particularly in the period leading up to the creation of the state of Israel, Zionist leaders did use the language of settler colonialism and often publicly discussed and adopted policies from other settler colonial states (see, e.g., Viterbo 2016). Zionist leaders presented the movement as a return to the promised homeland rather than emigration to a new land. This was not simply a discursive device to create a new nation out of nationally diverse Jewish migrants, it was an idea many Zionist leaders believed in, conveniently or otherwise (Herzl 1997). This meant that Zionist settlers, and after the creation of the state, Jewish immigrants and naturalised Israelis, have for most parts seen themselves as natives rather than settlers.

The perception of the Zionist settler as native is important to recognise because it underpins an unusual culinary process of self-indigenisation. The adoption and appropriation of regional and local food has been explained and justified as a process of reclamation; the above example of the olive tree demonstrates how food symbols become emblematic vehicles for Israelis to write themselves into the territory and history. It was and is argued, for example, that foods that were used by Jews in the Bible are by definition Jewish[6] or that adopting local foods was simply reclaiming Jewish heritage, as parts of the indigenous population were argued to be former Jews who adopted Islam in previous centuries (Mendel and Ranta 2014). As a result, through a slow but steady process, regional and local dishes and ingredients were adopted and appropriated into Israeli food culture, as local and Jewish, but mostly not as Arab or Palestinian (Mendel and Ranta 2014; and Raviv 2015); in many ways this slow but gradual process parallels the Zionist takeover of the land of Palestine, the dispossession of Palestinians, and the ongoing occupation of the Palestinian Territories.

Third, the process of Israeli self-indigenisation is further complicated by the historic presence of a small indigenous Jewish population and the emigration of large numbers of Jews from neighbouring Arab states, who, while not native to Palestine, are natives of the wider region. Tellingly, Jews from North Africa and the Middle East are often referred to in Israel as Mizrahi Jews (namely, Oriental Jews). While European Zionist settlers underwent a process of self-indigenising, the indigenous Jewish population undertook an inverse process of becoming settlers (Evri and Kotef 2022). The presence of local and migrating Jews from the region, many of whom were seen as Arab-Jews, further enabled the Zionist reclamation of regional and local food as Jewish; though it has also meant the appropriation of Mizrahi food as Israeli (see, e.g., Chaddad and Nizri 2019). The presence of Mizrahi Jews also shifted the balance from emulating European food cultures towards regional and local ones. It is important to note that this shift, which manifested in many cultural aspects, from

[6] See, for example, a discussion on the perceived biblical justification for Israeli ownership of hummus: Mendel and Ranta 2014; and the reclamation of ancient grape varieties: Monterescu and Handel 2019.

dress to music, occurred over time. Whereas initially Mizrahi food was presented at the tail end of cookbooks under the title of Jewish or ethnic food, today what is considered Israeli food is mostly regional, local, and Mizrahi food (there are clear overlaps between these categories). It is therefore important to recognise the multiple layers of what constitutes Israeli food vis-à-vis the Arab regional and Palestinian local, but also vis-à-vis Mirazhi food.

Lastly, while many settler colonial projects have either ended or have shifted towards decolonisation, in Israel, the colonial project is ongoing and at full force (Zreik 2016). It is ongoing as part of the Israeli occupation of the Palestinian Territories and the establishment, maintenance, and expansion of Jewish settlements at the expense of Palestinians. But also, through a continued process of self-indigenisation, which can be seen, for example, in the continued appropriation of Palestinian place names (Dahamshe 2021). In terms of food, this has meant a continued and even accelerated adoption and appropriation of regional and local dishes and ingredients and their recontextualisation as Israeli.

Cosmopolitan Appropriation: Depoliticising Food

Habasta restaurant and wine bar is located in the Karmel food market in Tel Aviv, Israel's economic centre and second largest city. Its name, which is Arabic in origin, literally means a market stall. The lowbrow name relates not only to the restaurant's location in a bustling market but also to its vibrant, noisy, and informal atmosphere and setting. It has consistently been listed as one of Israel's best restaurants and was even included in the 50 best restaurants in the world in the prestigious San Pellegrino list in 2019. According to the critics, the restaurant combines European cuisines and techniques with local Arab and Galilean influences, with a focus on terroir, seasonal, and local produce. In the menu, side by side Italian wines, mussels in white wine, and tiramisu, the restaurant serves dishes that are associated with Mizrahi, Arab, and Palestinian food cultures, such as *Chicken Maqlouba* (rice, vegetables, and chicken served upside down), *Chraime* (North African Jewish spicy fish), *Courgette in Yogurt with freekeh* (smoked cracked green wheat), and *khubeizah* and *Bubas* (mallow and lima beans). In many ways, the restaurant can be seen as the embodiment of a new patently urban Israeli food culture that has emerged over the past three decades. But what does this restaurant, and many like it, say about Israeli food culture?

Over the past three decades global and national food cultures have dramatically changed as a result of globalisation (Ranta and Ichijo 2016). On the one hand, globalisation has brought with it greater diffusion of culinary practices and ingredients, which has allowed consumers to experience a wider range of food cultures and products than ever before. On the other hand, globalisation has raised awareness and concerns over ethical, environmental, and cultural issues regarding food production, commodification, and consumption. This has renewed interest in local and organic food, artisanal production, slow foods, and foraging. These somewhat contradictory processes have contributed to the emergence of a mainstream foodie culture that seeks and celebrates diverse experiences, while at the same time values quality and authenticity (Johnston and Baumann 2015).

These trends have inspired Jewish-Israeli chefs and cookbook writers to re-examine the wider Middle Eastern and Mediterranean regions' culinary heritage; Israel's artisanal producers and local terroir; and the diverse culinary heritage of Israel's population, many of whom are second- or third-generation Jewish settlers-migrants. The result has been the evolution of a new concept of Israeli food culture, which restaurants such as *Habasta* epitomise—a food culture that is cosmopolitan and multicultural and 'borrows' freely from the best of all worlds and that to an extent acknowledges its various influences. Sometimes criticised as 'hipster' and 'trendy', this food culture no longer grapples with the tension between Europe and the region, but instead is malleable and incorporates both. For example, Israeli and international celebrity chef Asaf Granit, who gained fame in the Jerusalem restaurant *Machneyuda* (the name of the local food market), explained in an interview that his cooking was inspired by his 'Polish, German, Kurdish Jewish heritage' alongside 'local ethnic food' and global influences, 'like polenta and risotto' (Hadar 2019a). The new food culture can also be seen in the wide range of Israeli cookbooks that have been published internationally in recent years, such as *Jerusalem* (Ottolenghi and Tamimi 2012), *Zahav* (Solomonov 2015), and *The Book of New Israeli Food* (Gur 2007). These cookbooks, and many like them, acknowledge and celebrate Israel's diverse culinary cultures. They extol the virtues of different local food cultures, such as Jerusalemite or Galilean, and openly discuss the influence of Arab and Palestinian foods (Ranta and Prieto-Piastro 2019).

In the past, Palestinian chefs and food scholars have accused Jewish-Israeli chefs and food writers of cultural appropriation, of using Palestinian dishes and symbolic ingredients without acknowledging their Palestinian heritage. The appropriation of food was seen as analogous to the expropriation of land for the purpose of constructing and expanding the state of Israel, and later its settlements in the Occupied Territories. Palestinian food writer Reem Kassis told us she didn't 'think Palestinians have a problem with Israelis cooking their food or with there being Israeli restaurants, the problem comes with denying where these dishes have come from' (Van Leer Institute 2020). Chef Habib Daoud, a Palestinian citizen of Israel, emphasised this point by arguing that Israeli chefs should not deny the influence of Palestinian food culture and chefs (Van Leer Institute 2020).

The starting point of those who represent the new Israeli food culture, as *Habasta or Machneyuda* do, is the claim that they are not trying to produce an Israeli cuisine and are not denying the heritage of the dishes they produce. Indeed, those we interviewed stated clearly that they are not 'trying to make Israeli food', they are simply cooking 'without an agenda', cooking what they 'love and know'. They claim a cosmopolitan and inclusive food culture that embraces various influences, but is infused with their own cultural heritage and passion. They are open about using and, to an extent, being influenced by Arab and Palestinian food, chefs, food culture, and ingredients. Many spoke about how they were inspired by Palestinian chefs, such as Dukhul Safadi, whose restaurant Dianah in Nazareth has consistently been voted among the best in Israel (Goldstein 2017). According to them, their aim is not to appropriate but to simply produce good food. Their 'flirtation' with Arab and Palestinian food is based on their liberal conviction that local food, ingredients, and terroir (in Hebrew *Mekomiut*) can be shared.

This new willingness to acknowledge Arab and, to a lesser extent, Palestinian chefs, food, and influences is clearly of consequence. It has provided space for Palestinian chefs, amateur cooks, and food writers to showcase their food, restaurants, and culture to the Israeli public and audiences further afield, among others, in popular cooking shows on primetime TV, such as *MasterChef* (see, e.g., Grosglik and Lerner 2020; and Gvion 2017). It can also be seen as reproducing the Israeli image of an inclusive multicultural melting pot.

According to Palestinian chef Fadi Kattan, the unreflexive appropriation of Palestinian food, which he terms 'hipster restaurants', poses the greatest danger to Palestinian culinary heritage, precisely because of their cosmopolitan and pluralist image. In many ways, and following Kattan's point, Israeli chefs appear to exemplify Ghassan Hage's concept of cosmo-multiculturalism (Ghassan 1997). Israeli chefs and food writers are using and consuming Arab and Palestinian heritage, alongside global trends, to provide a culinary experience to their largely Jewish-Israeli and foreign clientele. All of the restaurants are in Jewish areas and the cookbooks are written in either Hebrew or English. This is similar to the commodification of indigenous food in other countries, which often comes under the guise of multiculturalism (Grey and Newman 2018). The main interest is in the produce and culinary heritage, but not in acknowledging the people who have produced it and their history of marginalisation.

While the growing visibility has been somewhat beneficial to Arab and Palestinian chefs and food writers, it has not necessarily led to greater inclusion or to changes in the uneven structures of power. For example, Arab and Palestinian chefs have struggled to attract Jewish clientele for their more upmarket restaurants and many Jewish-Israelis still associate Arab and Palestinian restaurants with cheap street food (Avieli 2017). Additionally, while extolling the virtues of using local produce and foraging, the same Jewish-Israeli chefs and food writers have largely remained silent over the rights of Palestinians in Israel and the West Bank to forage. Several Palestinian chefs have told us that while they are not allowed to forage for traditional plants, such as *'akub* and *white za'atar* (Hyssop or Syrian Oregano), as Israeli authorities claim these are endangered, they are sold openly in the main food market in Jerusalem (Eghbariah 2017). Another example is the growing use of *laban jamid* as a 'Palestinian

Parmesan' in Israeli restaurants, while the land rights of the Bedouin who mostly produce it are denied by the Israeli state.

The Jewish-Israeli chefs and food writers' desire to depoliticise their food is in many ways a political statement rather than a disavowal of politics; time and again we were told that the aim was to produce good food and not engage in politics. There is a clear recognition that any engagement with Palestinian issues and/or support for Palestinian rights comes with a price, a point made to us off the record several times.[7] Indeed, even the mentioning of Palestine and Palestinians is controversial and problematic, a point which we will discuss in greater detail below. Lastly, while some Jewish-Israeli chefs and food writers do acknowledge, if grudgingly and off the record, the problems faced by Palestinians and the problematic issues associated with using Arab and Palestinian food, many others do not. For example, a Jewish-Israeli chef told us that 'if I reinterpret an Arab dish surely that is not appropriation'. And, responding to a debate over the origins of *ara'is* (a popular Arab wedding dish of grilled pita stuffed with meat; the name comes from the Arabic word for bride), which has become popular in Israel, the Jewish-Israeli chef of a popular restaurant claimed that, though he acknowledges that the debate is political, 'we did not appropriate anything, we saw the dish in the Arab kitchen, and we gave it our version. In my restaurant I also do *chazeret* (a horseradish condiment used in the Passover meal and associated with Ashkenazi food), and beef tongue and also *ara'is*, and I am very proud of what I do' (Golan 2021).

Arab-Jewish Revival: Identity Politics in the Kitchen

'I am an Arab-Jew', 'I cook Arab cuisine and I am a proud Zionist', states Meir Adoni, one of Israel's leading chefs, in a recent interview (Hadar 2020). Not to be outdone, Guy Gamzo, an upcoming and trendy chef, declares that he is 'done with French and Italian cuisines': 'I am an Arab,

[7] Habasta, for instance, would not sell wines produced in West Bank settlements but does not declare this practice publicly.

and we are all Arabs. We all live in the Middle East, part of the same culture and eat the same foods' (Leshem 2020). These statements, and others like them voiced by chefs of North African and Middle Eastern Jewish descent like Adoni and Gamzo, have started to appear in Israeli media in recent years. They have appeared alongside the growing ascendance of Mizrahi food in Israel, which Avieli (2017) has termed the 'culinary Orientalization'. Many of the food items closely associated with Israeli food culture are of Mizrahi origin, from dishes such as *sabih* (an aubergine, egg, and salad sandwich) and *shakshuka* (eggs cooked in a spicy tomato sauce) to condiments such as *amba* (mango pickle)[8] and *schug* (a chilli condiment). In recent years there has been a rapid growth in restaurants and cookbooks focusing on the diversity of Mizrahi food cultures, including Iraqi, Moroccan, Turkish, Tunisian, Yemenite, and Persian Jewish food.

The re-emergence of the Arab-Jew and of Arab-Jewish/Mizrahi culture is a remarkable phenomenon. The term Arab-Jew has a controversial history. For most parts, Jews in the Arab world, in the long historical period before Zionism, participated in the social and cultural life of the region in Arabic. Shenhav (2006) writes about how the term was used predominantly by urban classes in the main cities in the early and middle period of the twentieth century, and by some of those who emigrated to Israel. Interestingly, the term was also used by the Ashkenazi-Zionist leadership as a pejorative to describe the Jews living in Palestine before Zionism and the Jews of the Arab world (Shenhav and Hever 2012).

Early Zionist accounts provide an indication to what a Zionist state was meant to resemble (see, e.g., Theodor Herzl's 1902 book *Altneuland* [Old-New Land]), with a clear focus on European Jews and European culture. It was only after World War II and the Holocaust, with the need to find more 'working and fighting hands' that the focus shifted to the Jewish population of the Middle East and North Africa (Segev 1998). The focus on European Jews and European culture was not missed on the new immigrants/settlers, who complained of being housed in worse conditions and provided with less support, and being lumped as Mizrahi Jews by the Ashkenazi leadership of the state, even though they came

[8] On *sabih* and *amba*, see Hart and Monterescu (2021).

from diverse countries and backgrounds. Their extensive food cultures were not always appreciated or seen as compatible with the newly created state. In the early period of the state many also felt like they were forced to learn to cook and eat like Ashkenazi Jews (Segev 1998).

Israel's approach to immigration was centred on the 'melting pot' ideology of assimilation, whereby the state's Ashkenazi leadership sought to transform the newly arriving Arab (Mizrahi) Jews into Israeli citizens. One of the main methods was deforming their Arab identity and culture and emphasising their Jewishness in opposition to their Arabness (Shenhav and Hever 2012). This could be seen in the push for new immigrants to speak Hebrew and in the stigma attached to speaking Arabic, the mother tongue of most Mizrahi Jews. As a result, most second- and third-generation Mizrahi Jews do not speak Arabic (Mendel 2014). These processes occurred side-by-side with direct government policies against the assimilation and integration of Palestinians, who were characterised as aggressive and uneducated (Mendel and Ranta 2014).

Taking the above points into account, the recent endorsement in Israel of Arab culinary heritage signals ironic poetic justice. Arabness has been transformed, at least in the culinary sphere, into a mark of quality and authenticity. The infamous 'Israeli salad' (chopped vegetable salad with olive oil and lemon juice) has been reframed over the past two decades as an 'Arab salad' (Ranta 2016). Hummus is now seen as something that Arabs and Palestinians do better (Hirsch 2011). To sell more food products some Jewish-Israeli companies have gone as far as labelling them as Arab (Ranta 2016). In terms of restaurants, the terms Mizrahi or Arab restaurant were used interchangeably to denote cheap and unsophisticated meals involving pita bread, salads, and grilled meat (Avieli 2017), but with the gourmetisation of Arab and Mizrahi food, this association has declined.

One interpretation of the positive transformation of the image of the Arab and the Arab-Jew, at least in the culinary field, is optimistic. Arab and Arab-Jewish food cultures are seen as more authentic, appealing, diverse, and more closely linked to the region, and its terroir. In many ways it is not surprising that Israel's food culture is based predominantly on Arab and Arab-Jewish food cultures. It also signifies a widening acceptance in Israel of its own diverse Arab (Jewish) population and its place in the Middle East and within the Arab world—an acceptance of sorts that Israel is no longer, in the words of a former prime minister, 'a villa in the jungle'.

There are, however, a number of issues that arise from this new Arab-Jewish revival discourse. To start with, this discourse involves not only the commodification, but also the appropriation of Arabness, often for commercial, at time for nationalistic purposes. It also further facilitates the reclamation of regional and local food as Jewish and therefore Israeli. For example, several Arab-Jewish chefs argue that there is little difference between regional Arab food, local Palestinian food, and Jewish Mizrahi food, so arguments over appropriation are void.

A more problematic issue arises from the endorsement and celebration of Arab food culture while refusing, at the same time, to acknowledge the existence of the Palestinian people, their food culture, and their rights. As argued by Palestinian lawyer Hicham Chabaita (Van Leer Institute 2020), there has been an Israeli tendency of misdirected acknowledgement, 'preferring to discuss the universal or regional Arab rather than the particular Arab or Palestinian living in Israel or the Occupied Territories'. Reem Kassis told us that 'from my experience, the very mention of the word Palestinian is enough for Israelis to say why are you turning this into something political. The "Arab" is easy to digest so long as he doesn't bring up the issue of the occupation and the injustice and imbalance of power'.

The unwillingness to mention the words Palestine and Palestinian are evident in most menus and interviews given by leading Jewish-Israeli chefs. With a few very notable exceptions,[9] Jewish-Israeli chefs are not comfortable referring to Palestinian food culture even when cooking distinguishable Palestinian dishes, such as *Knafeh Nabulsia* (sweet cheese pastry that is associated with the Palestinian city of Nablus). When discussing the use of local terroir and food cultures, Jewish-Israeli chefs are comfortable mentioning the terms Arab, Middle Eastern, and Levantine or referring to more perceived local regional food cultures, such as

[9] There are several Arab-Jewish chefs who do acknowledge these issues. Meir Adoni has spoken about how 'Arab-Palestinian food has always fertilised my culinary work' and that the food he cooks is inspired by his Jewish-Moroccan heritage and the Arab-Palestinian kitchen (Hadar 2020).

Jerusalemite or Galilean (a region in northern Israel/Palestine).[10] While many Jewish-Israeli chefs are happy to engage with local Palestinian food culture, they are not willing, for financial and political reasons, to acknowledge let alone engage with Palestinian history and politics.

The Reflexive Decoloniser: Recognition, Respect, and Terroir

A seasonal vegetable dish served with an olive oil vinaigrette does not appear at first a potential site for decolonisation. In many ways, Brut Bistro, a trendy Tel Aviv wine bar, which focuses on terroir, local produce, and seasonality and blends European classics with 'local Middle Eastern traditions' (Brut's website), is neither unusual nor an outlier in the Israeli culinary scene. However, by openly acknowledging the influence of Palestinian chefs and dishes and the use of Palestinian produce, Brut is creating a radical space for discussions over authenticity, indigeneity, and decolonisation in Israel/Palestine.

Questions concerning authenticity, appropriation, and indigeneity have started cropping up in studies of Israeli food culture over a decade ago (e.g., Hirsch 2011; and Melamed 2011). In many ways this is to be expected given the growing recognition of Arab and Palestinian food cultures and their influence on Israeli food culture. This influence was inevitable, as part of the Israeli process of self-indigenisation, but also because of the growing interest in artisanal produce and terroir. It is clear, though, that there is a difference between mentioning Arab and, to a lesser extent, Palestinian food cultures and influences and engaging with wider questions over indigeneity and settler colonialism. As we demonstrated above, many Jewish-Israeli chefs and food writers are and have been willing to mention and recognise Arab and, to a much lesser extent, Palestinian food cultures and influences. There have been, however, only a small number of chefs who have been willing to go beyond simply mentioning,

[10] Just to illustrate this point further, a well-known Jewish-Israeli chef who launched a 'Galilean Cooking School' and who celebrates his 'Galilean culinary identity' makes no mention of any Palestinian culinary influence.

to thinking more critically about the relationship between Israeli and Arab and Palestinian food and identity.

At one end of this more reflective approach are chefs that have moved beyond mere recognition to calling on Jewish-Israeli chefs to credit Palestinian chefs and support them. Israeli chef Tomer Agay argued in an interview that 'we need to give credit, [Israeli] chefs should say this is Palestinian *musakhan* (roast chicken on bread), why should we be embarrassed? In my restaurant I make sure to specify the Arabic name of the ingredients and where they come from. Courgette from Nablus, aubergines from Beit Safafa … when we are invited abroad to cook Israeli food, we should invite Arab chefs. Invite them to restaurant openings. You know what dishes you cook and who you have been influenced by' (Hadar 2019b).

Going beyond crediting and supporting to directly discussing and applying the settler colonial paradigm, however, is extremely rare and contentious. Some address the issue while highlighting its 'complexity' and refusing to attribute any primacy to one group or the other over local food. The depoliticised language of terroir or the story of food as a personal experience is apparent in the writing of Ottolenghi or Solomonov, for example. At the other end of the reflexive approach, there are a small number of Jewish-Israeli chefs that are reflexively and actively engaging in questions of appropriation, indigeneity, and decolonisation.

'The way Israelis engage with local food is colonialism proper', Yair Yossefi, the chef of Brut Bistro, tells us. Israeli chefs search for a 'local key to locality', 'but once the Israeli gets to Nazareth (a Palestinian city in Northern Israel) he does what his cousin does in the settlement of Tko'a (a settlement in the occupied West Bank known for its organic mushrooms)'. Yossefi is very critical of the way Israeli chefs engage with Palestinian food and Palestinian people and politics. He argues that the condescending attitude of Israeli chefs, who use terms such as 'Galilean food' rather than Arab or Palestinian food, prevents proper recognition by depoliticising the intimate relations between Israeli and Palestinian food. 'On the political Right we have chefs who say "my father is from Morocco so I can cook freekeh" (smoked green wheat) and even worse are those on the so-called [political] Left where people say "get politics out of the kitchen: politics is politics and food is food"'.

For Yossefi, the Israeli search for authenticity devoid of politics is a slippery slope towards normalisation. 'Hipster chefs join the global trend of foraging and localism, but the way they view the Palestinian Other is as colonialists', he says pessimistically. When Arab and Palestinian chefs are included in Israeli cooking shows, they are normally seen as 'pet Arabs' (*aravi mahmad*)—a concept used to denote paying lip service to political correctness by including a token 'good Arab'. According to Yossefi, for Israeli food culture to decolonise, 'there needs to be a recognition of the Other and more dialogue'. To address this challenge Yossefi lists on a blackboard at the centre of his restaurant the names and locations of his Palestinian suppliers. 'I tell my customers where my freekeh and lamb come from. I say that my cheese comes from Hebron. I tell the story and I go through the motions'. He is also active in bringing together Palestinian and Israeli chefs and food scholars to address these questions. His current project seeks to create a space for cooperation and exchange between figures such as Jenin-based Nasser Abufarha of Canaan Fair Trade and Palestinian chef Dukhul Safadi from Nazareth.

These initiatives however are few and far between considering the ongoing commodification of Palestinian food in Israel. 'Cooks don't make peace', admits Yossefi. 'The only thing that food talk can promote is a more respectful discourse'. While he is not optimistic about decolonising Israeli food, he still militates against the 'erasure' of Palestinian food and chefs. In his kitchen he produces food that seeks to reflect the local terroir, but it is precisely this attention to locality that makes him aware of culinary appropriation. Yossefi is one of the very few Jewish chefs in Israel that are reflexive and publicly vocal on the cultural politics of using and engaging with Palestinian food; 'I know no one else with whom I can talk to about these burning issues', he admits in our interview.

The situation is a bit different outside Israel, where recognition and acknowledgement of Palestinian people and issues exact less of a cost. Nevertheless, even outside Israel, Jewish-Israeli chefs are careful not to engage in questions of appropriation and colonialism. There are some exceptions to the rule, such as Gal Ben Moshe, the Michelin starred chef of Prism in Berlin, who produces 'Levantine cuisine, 4000 km away from the terroir I work with'. An outspoken critic of Israeli appropriation, Ben Moshe takes it further: 'in my opinion, the situation is even worse: it is not only about cultural appropriation but rather cultural humiliation. The za'atar and the 'akub are marked as endangered species not to protect them but to prevent access for the Palestinians who actually consume them. The Israeli couscous is another example. According to the Zionist discourse, the Israeli couscous (petitim), also known as the "Ben Gurion rice" was ostensibly Ben Gurion's (Israel's first Prime Minister) invention, but it's in fact a copy of the Palestinian maftul. We copied one of the bases of Palestinian food and marketed it as a major Israeli invention. It's part of the Occupation, of a cultural erasure. We sell ambience with Palestinian food and call it an Israeli restaurant'.

Conclusion: Food for Politics

A 2011 article in the Israeli newspaper *Ha'aretz* (Melamed 2011) celebrated the 'penetration' of Arab and Palestinian food into the heart of Tel Aviv's food scene. The article went on to argue that 'the leading Israeli chefs are more Arab than ever' and that 'it appears the combination and mixing of the kitchens is part of a natural process of assimilation of the two nations living together. A combination that is inevitable given that the kitchen is an inseparable part of any culture'. The article also quoted a chef who argued that 'our connection to the Palestinian kitchen is the most natural and required given the reality of the two nations, and, in my view, the kitchen that is being created is amazing'. Ten years later, there is no doubt that Arab and Palestinian food cultures have become prominent features in Israeli food culture. On popular food blogs and on social media, Jewish-Israeli foodies argue where to find the best Arab hummus or *knafeh* and how to cook vegan *freekeh*. It is also clear that there is growing recognition of the influence of these food cultures on Israeli food culture.

The questions we started with continue to challenge our political critique: is Israeli food culture undergoing a paradigmatic shift? Are we seeing signs of the decolonisation of Israeli food? It is premature to provide a straightforward answer to these questions. To start with, the answers are complicated by the murky boundary that exists between culinary appropriation and conscious recognition, which may just be the difference between different discursive framings of the same dish. Also, culinary diffusion is part of the culinary interactions that occur on a regular basis globally, as food cultures adopt, adapt, and change. It is clear that we are witnessing change in Israeli food culture regarding the role and place of Arab and Palestinian food culture and people. However, it is also clear that much of this change is occurring within hegemonic Zionist parameters that delimit and delineate the scope and nature of this change. For the most part, Jewish-Israeli chefs and food writers control the terms of these culinary interactions. Moreover, the ambivalent relationship with the food of the political Other takes part within larger processes: the normalisation of Israeli expansion into the West Bank, the rise of settler organic farmers, and the strategic blurring of the Green Line.[11] Most of the Jewish-Israeli chefs and food writers, who recognise and acknowledge Arab and Palestinian influences on Israeli food, talk about the importance of mutual respect, about supporting Arab and Palestinian chefs and producers and about coexistence. What most are not talking about, or are saying very quietly, are the ongoing occupation of the Palestinian Territories and the daily marginalisation facing Palestinians living in Israel.

For the Palestinian chefs and food writers we spoke to, the decolonisation of Israeli food means first and foremost the recognition of the place and role of Arab and Palestinian food in Israel and the end to cultural appropriation. The next step is for Israeli chefs to call for an end to the occupation and the marginalisation and oppression of Palestinians, particularly in the realm of food. As Kassis told us: 'while I laud the efforts of these [Israeli] chefs to recognise the Palestinian origins and to use Palestinian suppliers, what good does that serve when the political environment prevents those very Palestinians from enjoying a normal quality of life? When a Palestinian farmer is not able to access his land and yet a settler is able to come make use of that land and put a swimming pool on it and allow his kids to swim in it while the rightful Palestinian owner cannot even water his crops, how do any of those small steps taken by Israeli chefs move the needle?'

[11] The Green Line is the name given to the 1948 ceasefire lines that demarcate the political boundary between Israel and the occupied West Bank.

The point made by Kassis is one that we have encountered on many occasions before: how can small steps taken regarding food issues make any difference?[12] This argument has several layers to it. On the one hand it is argued that small steps taken by Jewish-Israeli chefs and food writers can't in and of themselves make much of a difference. On the other hand, there is the argument that food is not a political and/or an important issue, a point made to us by several Jewish-Israeli chefs who asked for politics to be left out of the kitchen. In other words, whether there is some observable change in Israeli food is irrelevant to wider debates on Israeli-Palestinian relations and the occupation.

These points are off the mark. It is precisely because food is seen as apolitical, socially shared, and culturally inclusive, that it is an important tool, as evidenced by the growing interest triggered by the publications of Palestinian cookbooks, such as *Falastin* (Tamimi and Wigely 2020), *The Palestinian Table* (Kassis 2017), and *The Gaza Kitchen* (El-Haddad and Schmitt 2013). It can be argued that for lack of other venues of dialogue, Israeli engagement with Palestinian food marks it as one of the few areas with the potential to serve as a starting point to at least think of decolonisation (a point made by chefs such as Yossefi and Ben Moshe). The proliferation of new Palestinian restaurants which frame themselves as such is a major turning point paradoxically made possible by culinary appropriation (in Jaffa alone we witness three such new food projects). The language of the right to terroir and gastronomic heritage can thus lead either to further recognition or to the depoliticisation and commodification of the culinary field. In the typology we offered, the cosmopolitan chefs consider politics as a hot potato, while the Jewish-Arab chefs ethnicise gastronomy mostly in Jewish terms. At this stage, only the reflexive decoloniser is willing to critically introduce politics into the kitchen.

[12] While we were often told unequivocally that hummus and falafel cannot bring about peace, a graffiti by Banksy on the Separation Wall in Bethlehem reads: Make Hummus Not War!

Bibliography

Abufarha, Nasser. 2008. Land of Symbols: Cactus, Poppies, Orange and Olive Trees in Palestine, Identities. *Global Studies in Culture and Power* 15 (3): 343–368.

Almog, Oz. 2000. *The Sabra: The Creation of the New Jew*. Berkeley, CA: University of California Press.

Avieli, Nir. 2017. *Food and Power: A Culinary Ethnography of Israel*. Berkeley: University of California Press.

Bardenstein, Carol. 1998. Threads of memory and Discourses of Rootedness: of Trees, Oranges and the Prickly-Pear Cactus in Israel/Palestine. *Edebiyat* 8: 1–36.

Chaddad, Rafram, and Yigal Nizri. 2019. How Shakshuka and Other Middle Eastern Dishes Turned into Iconic 'Jewish Food'. *Haartez*, November 21. https://www.haaretz.com/food/.premium.MAGAZINE-how-shakshuka-and-other-arab-and-mideastern-dishes-became-iconic-jewish-food-1.8161982. [in Hebrew].

Dahamshe, Amer. 2021. Palestinian Arabic versus Israeli Hebrew Place-Names: Comparative Cultural Reading of Landscape Nomenclature and Israeli Renaming Strategies. *Journal of Holy Land and Palestine Studies* 20(1): 62–82. https://doi.org/10.3366/hlps.2021.0258

Eghbariah, Rabea. 2017. The Criminalization of Za'atar and Akkoub: On Edible Plants Palestinian Cuisine and Israeli Plant Protection Laws. In *Studies in Food Law*, ed. Yofi Tirosh and Aeyal Gross, 492–533. [in Hebrew]

Elgenius, Gabriella. 2011. *Symbols of Nations and Nationalism: Celebrating Nationhood*. Basingstoke: Palgrave Macmillan.

El-Haddad, Laila, and Maggie Schmitt. 2013. *The Gaza Kitchen: A Palestinian Culinary Journey*. Charlottesville: Just World Books.

Evri, Yuval, and Hagar Kotef. 2022. When does a native become a settler? (With apologies to Zreik and Mamdani). *Constellations* 29: 3–8. https://doi.org/10.1111/1467-8675.12470

Ghassan, Hage. 1997. At Home in the Entrails of the West: Multiculturalism, Ethnic Food and Migrant Home-Building. In *Home/World: Space, Community and Marginality in Sydney's West*, ed. Helen Grace et al., 99–153. Annandale, NSW: Pluto Press.

Golan, Tiki. 2021. From Where and How did the Arais Make Its Way to the Menus of the Israeli Meat Restaurants. April 16. https://www.ynet.co.il/food/foodnews/article/rkIQQT8Id. [in Hebrew].

Goldstein, Rita. 2017. The Best Restaurants in Israel: All the Winners. *Mako*, September 17. https://www.mako.co.il/food-restaurants/best-israels-best-restaurants/Article-a09d721fb9f7e51006.htm?sCh=f7e52b939905d510&pId=885471177. [in Hebrew].

Grey, Sam, and Lenore Newman. 2018. Beyond Culinary Colonialism: Indigenous Food Sovereignty, Liberal Multiculturalism, and the Control of Gastronomic Capital. *Agriculture and Human Values* 35: 717–730.

Grosglik, Rafi. 2017. *Organic Food in Israel: Resistance, Assimilation and Global Culture*. Sderot: Resling. [in Hebrew].

———. 2021. *Globalizing Organic: Nationalism, Neoliberalism, and Alternative Food in Israel*. New York: SUNY Press.

Grosglik, Rafi, and Julia Lerner. 2020. Gastro-emotivism: How MasterChef Israel Produces Therapeutic Collective Belongings. *European Journal of Cultural Studies*. https://doi.org/10.1177/1367549420902801.

Grosglik, Rafi, Ariel Handel, and Daniel Monterescu. 2021. Soil, territory, land: The spatial politics of settler organic farming in the West Bank. *Environment and Planning D: Society and Space*. https://doi.org/10.1177/02637758211041121.

Gur, Janna. 2007. *The Book of New Israeli Food: A Culinary Journey*. New York: Schocken Books.

Gutkowski, N., D. Disegni, and D. Rabinowitz. 2013. Fair Trade Olive Oil and Its Environmental Impact. *The Journal of Ecology and the Environment* 4 (1): 22–13.

Gvion, Liora. 2017. 'From an Arab I would have Expected…': 'Masterchef', Hegemony and Exclusion. In *Studies in Food Law*, ed. Yofi Tirosh and Aeyal Gross, 561–587. [in Hebrew].

Hadar, Alon. 2019a. Chef Asaf Granit: 'Success is Comprised of many Elements that most Restaurateurs don't Consistently Do'. *Hashulchan*, April 2. https://www.hashulchan.co.il/%D7%A8%D7%99%D7%90%D7%99%D7%95%D7%9F-%D7%A2%D7%9D-%D7%A9%D7%A3-%D7%90%D7%A1%D7%A3-%D7%92%D7%A8%D7%A0%D7%99%D7%98/. [in Hebrew].

———. 2019b. The Israeli Chefs did not Invent Anything New. *Yediot*, October 5. https://www.yediot.co.il/articles/0,7340,L-5602948,00.html?fbclid=IwAR0iFEY28nveKpE9iocLk-DsStQ5PJR4F-5XSVzeW6bqyK0iEI5wzH3XPfQ. [in Hebrew].

———. 2020. Meir Adoni: 'I cook Arab Food and I am a Proud Zionist'. *Hashulchan*, April 5. https://www.hashulchan.co.il/%d7%a8%d7%99%d

7%90%d7%99%d7%95%d7%9f-%d7%a2%d7%9d-%d7%a9%d7%a3-%d7%9e%d7%90%d7%99%d7%a8-%d7%90%d7%93%d7%95%d7%a0%d7%99/. [in Hebrew].

Hart, Joel, and Daniel Monterescu. 2021. The Sensorial Life of Amba: Taste, Smell and Culinary Nostalgia for Iraqi Jews in London and Israel. In *Food, the Senses and the City*, ed. Ferne Edwards et al., 95–107. London: Routledge.

Herzl, Theodor. 1997. *Altneuland—The Old-New Land [1902]*. Tel Aviv: Babel Publishers. [in Hebrew].

Hirsch, Dafna. 2011. Hummus Is Best When It Is Fresh and Made by Arabs: The Gourmetization of Hummus in Israel and the Return of the Repressed Arab. *American Ethnologist* 38 (4): 617–630.

Johnston, J., and S. Baumann. 2015. *Foodies: Democracy and Distinction in the Gourmet Foodscape*. Abingdon: Routledge.

Kassis, Reem. 2017. *The Palestinian Table*. London: Phaidon.

Leshem, Eitan. 2020. 'I am done with French and Italian cooking', the Arab Revolution of Guy Gamzo and the Aria Restaurant. *Haaretz*, June 3. https://www.haaretz.co.il/food/food-news/.premium-MAGAZINE-1.8887105. [in Hebrew].

Melamed, Dana. 2011. Al-Yehud Kitchen: Arab Food has Conquered the Leading Chefs. *Haaretz*, February 23. https://www.haaretz.co.il/food/1.3272775. [in Hebrew].

Mendel, Yonatan. 2014. *The Creation of Israeli Arabic: Security and Politics in Arabic Studies in Israel, Palgrave Studies in Languages at War*. London: Palgrave Macmillan.

Mendel, Yonatan, and Ronald Ranta. 2014. Consuming Palestine: Palestine and Palestinians in Israeli Food Culture. *Ethnicities* 14 (3): 412–435.

Meneley, Anne. 2011. Blood, Sweat and Tears in a Bottle of Palestinian Extra-Virgin Olive Oil. *Food, Culture & Society* 14 (2): 275–292.

Monterescu, Daniel, and Ariel Handel. 2019. Liquid Indigeneity: Wine, Science, and Colonial Politics in Israel/Palestine. *American Ethnologist* 46 (3): 313–327.

Ottolenghi, Yotam, and S. Tamimi. 2012. *Jerusalem*. London: Ebury Press.

Prieto-Piastro, C.R. 2017. *Teach Me How to Eat as an Israeli: Nation, Food Culture and Identity in Israel*. PhD dissertation, King's College London.

Ranta, Ronald. 2016. Re-Arabizing Israeli Food Culture. *Food, Culture & Society* 18 (4): 611–627.

Ranta, Ronald, and Atsuko Ichijo. 2016. *Food, National identity and Nationalism: From the Everyday to the Global*. London: Palgrave Macmillan.

Ranta, Ronald, and Claudia Prieto-Piastro. 2019. Does Israeli Food Exist? The Multifaceted and Complex Masking of National Food. In Ranta et al. (eds), *The Emergence of National Food*. London: Bloomsbury.

Ranta, Ronald, and Yonatan Mendel. 2016. *From the Arab Other to the Israeli Self: Palestinian Culture in the Making of Israeli National Identity*. London: Routledge.

Raviv, Yael. 2015. *Falafel Nation*. Lincoln, NE and London: University of Nebraska Press.

Raz-Krakotzkin, Amnon. 1993. Exile Within Sovereignty: Towards a Critique of the 'Negation of Exile' in Israeli Culture'. *Theory and Criticism* 4: 23–56. [in Hebrew].

Roden, Claudia. 1999. *The Book of Jewish Food: An Odyssey from Samarkand and Vilna to the Present Day*. London: Penguin.

Said, Edward. 1979. Zionism from the Standpoint of Its Victims. *Social Text* 1 (Winter): 7–58.

Segev, Tom. 1998. *1949: The First Israelis*. New York: Owl Books.

Shafir, Gershon. 1989. *Land, Labour, and the Origins of the Israeli–Palestinian Conflict, 1882–1914*. Cambridge: Cambridge University Press.

Shenhav, Yehouda. 2006. *The Arab Jews: A Postcolonial Reading of Nationalism, Religion, and Ethnicity*. Stanford: Stanford University Press.

Shenhav, Yehouda, and Hannan Hever. 2012. 'Arab Jews' After Structuralism: Zionist Discourse and the (de)formation of an Ethnic Identity. *Social Identities* 18 (1): 101–118.

Solomonov, Michael. 2015. *Zahav: A World of Israeli Cooking*. Boston, New York: Houghton Mifflin Harcourt Publishing Company.

Tamimi, Sami, and Tara Wigely. 2020. *Falastin: A Cookbook*. London: Ebury Press.

Van Leer Institute. (2020). Turning the Tables: Palestinian and Israeli Food on the Cutting Board.

Veracini, Lorenzo. 2010. *Settler Colonialism: A Theoretical Overview*. Basingstoke: Palgrave Macmillan.

Viterbo, Hedi. 2016. Ties of Separation: Analogy and Generational Segregation in North America, Australia, and Israel/Palestine. *Brooklyn Journal of International Law* 42: 696–760.

Zaban, Yair. 2016. *Land of Milk and Hummus: A Study of Israeli Culinary Culture*. Ramat Gan: Afik Books. [in Hebrew].

Zreik, Raef. 2016. When Does a Settler Become a Native? (with Apologies to Mamdani). *Constellations* 23 (1): 351–364.

9

'Uneasy Lies the Head That Wears the Crown': Lamb or Kangaroo, Which Should Reign Supreme? The Implications of Heroising a Settler Colonial Food Icon as National Identity

Jacqueline Newling

Advance Australian Fare

At the 18th Symposium of Australian Gastronomy in 2011, I put forward a call to action for kangaroo to replace lamb as the celebratory meat on Australia Day (Newling 2011a, 116–117). This was in response to a national advertising campaign developed by Meat and Livestock Australia (MLA), which had claimed for several years that it was 'un-Australian' not to eat lamb on Australia Day (Bennett 2018; Tibbertsma 2017; Effie Australia 2009). As a gastronomic historian, I challenged this

---

This work was prepared on Wangal Country. I pay my respects to Australia's First Peoples, their Elders past and present, and acknowledge their deep and ongoing connection to their ancestral lands.

J. Newling (✉)
Honorary Associate in History, The University of Sydney, Camperdown, NSW, Australia

R. Ranta et al. (eds.), *'Going Native?'*, Food and Identity in a Globalising World,
https://doi.org/10.1007/978-3-030-96268-5_9

concept—why should lamb wear the crown when it is a powerful symbol of imperialism and settler colonialism; surely, a native alternative such as kangaroo would be more appropriate. In the decade since, I have been reconsidering this stance.

Australia Day is celebrated on January 26, marking the date of the official founding of the colony in 1788. The anniversary of British colonisation has been celebrated in various ways since 1789, but January 26 was declared the nationwide Australia Day public holiday in 1994.[1] Australia is in an ongoing process of increased awareness and confrontation with the 'uncomfortable truths' of its gross injustices towards Aboriginal people in the past and present, from social subjugation to genocide.[2] Celebrating January 26 has long been recognised as an insult to Indigenous Australians, who in 1938 (upon the 150th anniversary of British colonisation) proclaimed the day as a 'Day of Mourning' (AIATSIS n.d.; Australian Human Rights Commission 2020). The date is referred to by many Indigenous and non-Indigenous Australians as Invasion Day, raising questions regarding the morality and ethics of celebrating a day that marks the unlawful British occupation and colonisation of Australia, which led to dispossessing Aboriginal peoples—often violently—of their lands and culture across Australia. Respecting this cultural sensitivity, there has been, in recent years, an active campaign to 'Change the Date' of the Australia Day celebration. Although greatly contested, the concept that has been embraced by many individuals and organisations, cancelling or rescheduling celebrations and events once held on or aligned with the holiday (Costantoura 2020). Other recent initiatives acknowledging Indigenous Australians' ancient and continuing presence and culture include changing the words in the national anthem, Advance Australia Fair, from 'we are *young* and free' to 'we are *one* and free'. More significantly, First Peoples community leaders are negotiating with government for constitutional change to ensure Indigenous people a voice to parliament, as well as other measures of recognition and empowerment. The government's response is still developing, with no resolution yet, that

[1] For discussion regarding Australia Day as a nationalising political construct, see McKenna (2012).

[2] This is not necessarily a new phenomenon. Social commentators have acknowledged this in varying degrees of severity in the past. For example, Mundy (2006, 224, 226); Barnard (1938).

honours Australia's First Nations people's ambition to have their voice enshrined in the Constitution.[3]

In the midst of this slowly growing national consciousness, this chapter considers Australia's culinary identity in the contexts of the nation's settler colonial heritage, its multicultural diversity and acknowledgement of the dispossession of this country's First Peoples. Drawing on historical references, period cookery texts and current scholarship concerning Australia's relationship with native foods, particularly kangaroo, this study argues that presenting lamb as the national celebratory meat supports a broader legacy of settler colonial self-indigenisation and cultural 'whitewashing'. It then discusses the moral implications of adopting kangaroo as an alternative celebratory and symbolic meat, albeit a more environmentally responsible choice, given the current unresolved socio-political issues of sovereignty.

Anthropologist Claude Levi-Strauss famously stated that a society's cookery is a language into which its structure is translated, but that this culinary voice can also reveal its contradictions (1990, 495). In a similar vein, Australian culinary historian Barbara Santich notes that a society's food represents its myths and mores, its practices and priorities (1996, 43). With the exception of seafood, native foods have generally occupied a marginal place in Australia's settler colonial cookery culture or, at times, been completely absent. The status of native foods from 1788 to the present has been the subject of much interpretation and scholarship across various disciplines. Some scholars and commentators see the underrepresentation of native foods in the mainstream culinary repertoire as a reflection of non-Indigenous Australians' tenuous relationship with the land and environment and their ignorance of or dismissive attitudes towards First Nations people's knowledge and epistemologies (Pascoe 2016, 88–89). Some also posit that resistance to native produce stems from guilt associated with the maltreatment of Aboriginal people and illegal dispossession of their land (Craw 2012, 19–21), while others see it as evidence of disrespect for and prejudice towards Australia's First Nations people or, put more bluntly, 'food racism' (Newton 2014, 81, 244, 262; 2016, 91, 110).

[3] See McKay (2017) and Aboriginal Land Council of New South Wales (2021).

Australians' willingness to eat kangaroo meat, in particular, has gained scholars' attention. The prevalence of lamb (or mutton in the colonial era, 1788–1901) on Australia's tables, historically and contemporarily, provides fertile ground for analysis of the way settler colonialism is translated onto the plate. In my mind, the success of a marketing campaign promulgating lamb as the nation's meat of choice on Australia Day demonstrates, albeit in culinary terms, the extent to which structures of settler colonialism are 'relentlessly active in the present' (Cavanaugh 2012, 31). Promoting an introduced animal—one integral to the geographical, political and economic structure of colonial expansion and with a detrimental effect on the environment and Aboriginal people's traditional homelands and culture—suggests cultural arrogance and insensitivity. It illuminates the contradictions between Australia's perception of itself as an independent, multicultural, 'fair go for all' post-colonial nation and the unsettling anxieties of reconciling past and ongoing wrongdoings and injustices inflicted upon Indigenous Australians.

Mongrels and Mavericks

Australia has a well-stocked 'tucker bag' of foods that are staunchly defended as national icons. Among them are Vegemite, Lamington cakes, Anzac biscuits, meat pies and backyard 'barbies' (barbeques), but few of them are uniquely Australian, having comparable counterparts in other countries. Lamb, either roasted or barbequed, is also identified as an entrenched Australian food icon (Buresti 2018, 6; Greenwood 2011, 61). The absence of a distinguishable cuisine, let alone one that celebrates native Australian produce, is a matter of intense discussion, and in some circles, concern, about Australian society and the nation's cultural identity (Bannerman 2011; Greenwood 2011; Newton, 2018; Symons 2007). Since its colonial beginnings in 1788, the nation's diet has reflected its Anglo-Celtic settler colonial heritage, progressing from convict rations to the 'mutton, tea and damper' that embodied the pioneering spirit of 'bushmen' settler colonists of the nineteenth century[4] to a more urbane but reputedly unimaginative 'meat and three veg' that dominated the menu in many households in the twentieth century.[5] The influences of post–Second World War migration and affordable air travel since the 1970s have been attributed to the increase in Australians' exposure to other cultures' foods. The adoption of elements of others' cuisines has resulted in Australia having a 'mongrel cuisine' (Newton 2014, 152–154, 248)—one of 'migrants and mavericks' (Greenwood 2011, 63). Rather than having a mono-cuisine 'built on centuries of racial continuity' food scholar Helen Greenwood says 'ours is a poly-cuisine' which reflects Australia's social makeup as an 'assembly of transplanted nationalities' (2011, 63–64), or in Colin Bannerman's words, 'cultural jetsam' (2019, 82). By the 1990s the exotic was 'routinised' and commonplace (Foley 2005, 27). 'Spag bol'—spaghetti bolognaise, an Australianised version of *tagliatelle al ragu*—and similarly reductive Asian curries or stir-fries are just as likely to emerge from the Australian kitchen as conventional Anglo-Celtic roasts, grills, stews and casseroles (Dale 2020). Australia's 'promiscuous multiculinarity' (Newton 2014, 258) exemplifies the societal drift from its British and settler colonial roots to a country that perceives itself as globally connected, culturally diverse, inclusive and accepting of others and their tastes. But while consuming foods of the Other may promote multiculturalism as an ideal and help to develop an understanding of others' cultures, it is not proof of anti-racism (Anderson and Benbow 2015, 34–43; Probyn 2011, 2; Flowers and Swan 2012, 2–3). Similarly, Lorenzo Veracini (2011, 7–8), following Mahmood Mamdani, reminds us that 'privileging the indigenous over the non-indigenous' may turn colonial logic upside down, but it does not necessarily change it.

[4] The phrase was immortalised in a poem by Francis Lancelott (1852, 263–264) that describes 'Bush cuisine' as variations of mutton, tea and damper, seven days a week.

[5] This is not to say that Australian food was boring or lacking in flavour. For discussion, see Bannerman (2019); Newton (2018); O'Brien (2016); Santich (1996); Santich (2012).

Indigenising the Menu

Historically, native foods played an important role in the colonial diet, adding freshness and variety to salted meats (essential before artificial refrigeration) and providing local alternatives to English game. For various reasons (see Newling 2011b, 39–42, 47) they were not exploited commercially, but there is ample evidence that local foods were welcome additions to the diet in the early days of colonisation and throughout the nineteenth century (Bannerman 2019, 33; Newling 2011b, 27–48; Santich 2011, 65–78; Singley 2012, 27–42).

Native foods were invariably compared with English and European 'equivalents' and cooked accordingly. Kangaroo, which is extremely lean, was likened to venison, and wallaby to hare. Their tails were cooked the way Britons would ox-tail. The 'kangaroo steamer'—fatty salt-pork combined with slices of kangaroo simmered in a sealed jar with whatever aromatics were at hand—was a derivation of English jugged hare (Santich 1996, 118–120). Wallaby was also 'jugged' and at one point exported to England as a potential commercial product. 'Slippery bob'—kangaroo brains fried in emu fat (Abbott 1864)—is popularly cited as novel, but not dissimilar to any other fried brains dish. Local fish were stuffed and baked or boiled and served with oyster sauce. Parrots went into pies, wattle-birds (and almost every other game meat) were curried, and wonga-wonga pigeon, valued for its plump white breast meat, was served with bread sauce, a traditional accompaniment to fowl or pheasant (Pearson 1894, Maclurcan 1905). Kangaroo and wonga-wonga pigeon were recommended for acclimatisation in Europe—they translated easily to familiar applications (Granville et al. 1860, 24–25; Singley 2012, 37).

Queensland hotelier and cookery author, Hannah Maclurcan, boasted that she often served native game birds and animals in the place of more conventional (for the times) English species. Passing off wallaby for hare and scrub turkey for pheasant, she claimed 'no one has been any the wiser' (Maclurcan 1905, 206, 214). The fact that she disguised them as more readily acceptable northern hemisphere species for her dining guests suggests market resistance to local alternatives. Nonetheless, she openly admits her deception in her self-titled cookbook, *Mrs Maclurcan's Cookery Book* which contains recipes *Specially Suitable for Australia*, including many for native fish, game birds and animals.

Fellow Queenslander Wilhelmina Rawson, author of *The Antipodean Cookery Book and Kitchen Companion* (1895), acknowledges local Aboriginal people for her knowledge of native foods and bush medicines (1992, 54. See also Wessell's chapter in this volume, 109–126). Rawson assures readers that 'there is a great amount of pleasure' in experimenting with 'primitive materials' (55). She insists 'there is nothing nasty or disgusting' about the 'white wood grub which the blacks are fond of' and highly recommends the tail of a young iguana, 'cooked black's fashion in the ashes' or 'cut up and curried' (54, 109). She concedes, however, that 'ground game' ('wallaby, paddy-melon, bandicoot, iguana etc.') had a 'strong peculiar flavour', but soaking the carcass in a vinegar solution or 'burying it in the ground for several hours' could ameliorate this (108–109). The preparation process—skinning, gutting, washing and cutting up the animal—was, however, an 'ordeal' and 'distasteful'. The coarse nature of these foods and the interventions required to make them palatable play to 'settler discourses of perversity, deficiency, inferiority' associated with primitivism (Instone 2005, 139, 140. See also Mabli's chapter in this volume, 47–64). They reinforce the view that Australia is a wild place requiring taming and order—in other words, civilising (Craw 2012, 19). Making wild resources edible was one of 'the first steps of a colonial project that involved the control and domination of the natural environment' (Singley 2012, 30). Applying sophisticated cooking applications to ameliorate their coarse primitive state was a mark of colonial triumph over nature (Instone 2005, 136, 137). By co-opting these native food sources into Eurocentric cuisine, Rawson and Maclurcan typify the concept of the 'new native' (Youé 2018, 70), claiming authenticity and authority over Aboriginal peoples' traditional foods by effectively appropriating them for their own use.

Standing in contrast to these cookbooks is the *Art of Living in Australia* (1893) by Melbourne-based physician Philip Muskett. Muskett urged Australians to adopt a lighter, less meat-driven Mediterranean-style diet, better suited to local soils and climate. Apart from fish and shellfish there is no mention of native foods and the 300 recipes written to accompany his treatise (by English-trained cookery teacher, Harriet Wicken) are decidedly conventional. Native produce was no doubt more visible and accessible for women in northern Queensland with hands-on experience in the kitchen than for a gentleman doctor residing in the metropole.

Recipes drawing on native foods other than seafood are rare in cookbooks published from the early 1900s. Kangaroo and wallaby are the most typically included, with the occasional steak, mince and 'steamed' dish, but their tails were still popular for soups. Some books offer recipes for emu's and swan's eggs (Watkie 1949) and although there are myriad edible plants (see Wessell's chapter in this volume, 109–126), few beyond quandong and the wild hibiscus 'rosella' make an appearance (Pearson, 1894; Rutledge 1904, 1937).

Australia's colonial era officially ended with Federation on January 1, 1901, following a decade-long campaign propelled with intense nationalistic sentiment (Summers 2000, np). Underpinned by a strong commitment to Crown and Empire and a fear of non-British migrant workers, the Immigration Restriction Act, otherwise known as the White Australia Policy, was legislated in December 1901. Subsequent parliamentary debate in 1902 about First Nations people's democratic rights indicates the sense of superiority and prejudice that many—but, importantly, not all—white Australians carried towards Aboriginal people (Summers 2000, np).

While Australians were happy to include exotic sauces, relishes and spices in their diets (especially the then-ubiquitous curry powder, naturalised in both Britain and Australia by the mid-1800s), the diet reflected the largely Anglo-Celtic face of Australia. French cuisine was regarded as sophisticated, and French-style dishes were served on formal or refined tables, but distrust and dislike of people from other cultures may explain the lack of foreign dishes and similarly, native foods, in the Australian culinary repertoire in the early twentieth century.[6]

Kangaroo Consumption

The majority of native land animals and birds are now protected by law, but emu, crocodile, wallaby and kangaroo can be harvested under licence. Australian native foods are often referred to as 'bush' foods, or more casually 'bush tucker' and 'bush tukka', inherently suggesting they are found and eaten in remote 'outback' areas as survival foods or by people who

[6] For discussion, see Bannerman (2019); Newton (2018); O'Brien (2016); Santich (1996); Santich (2012).

understand 'bush lore' (Instone 2005, 137). Their place in the mainstream Australian diet today remains niche, often as novel fare for tourists. Wallaby, emu and crocodile can be found on tourist menus and specialist game providers, but only kangaroo meat is readily available in supermarkets.

Australian consumers have had a complex relationship with kangaroo meat. Although welcomed onto many colonial tables, by the late 1800s kangaroos and wallabies were regarded as pests, as they competed with farmed livestock for green pick, especially in periods of drought. Under the Pastures and Stock Protection Act of 1880, bounties were paid for scalps of these and other 'noxious animals' in the sheep districts of New South Wales.[7] Harvesters could profit from the animals' hides, but there was little appetite for converting wild-caught kangaroo carcasses into commercialised table meat. Investment in artificial refrigeration infrastructure was reserved for animals that could be driven along stock routes to centralised abattoirs which were linked to the cold-chain systems that would transport lamb and beef to metropolitan markets. As a result, urban consumers had little opportunity to become familiar with cooking and eating kangaroo meat.

Commercial interest in kangaroo meat grew from the late 1950s as an export commodity—for foreign tables as exotic game, but also for the burgeoning American canned pet food industry. Exports ceased during the 1970s, to allow now drastically depleted stocks to recover. Although harvesting was reactivated in 1980 for certain species, on percentage per-capita bases, the sale of kangaroo meat for human consumption was banned in all but one state, in part to protect the market share of beef and lamb. Further changes to the law allowed kangaroo meat to be sold as table meat throughout the nation in 1993, but the association with pet food and public health scares about the hygiene quality of wild-caught meat did little for the reputation of kangaroo among domestic consumers (Pople and Grigg 1999b).[8] Adventurous and 'open-minded' chefs, how-

[7] Rabbits were also listed as noxious animals in the Act, having become feral after their release into the wild in the late 1850s. Once perfectly acceptable, they became associated with poverty much maligned as paupers' food.

[8] Four species are approved for commercial harvesting, with state-based management plans monitoring population numbers and setting harvesting quotas, typically 10–15% of the total population, which stands around 50 million (Macro Meats 2020).

ever, looking for variety and difference on their menus, helped elevate the status of kangaroo from bush tucker, outback game or gimmicky tourist tantaliser to gourmet fare (Alexander 1996, 357–358; Appleby 2010, 57–59, 90; Probyn 2011, 35, 38–39).

Normalising kangaroo as a table meat has been challenging for many industry 'pioneers' who have had to battle against cultural and emotional barriers to kangaroo consumption, including the symbolic nature of kangaroo as an Australian icon. A kangaroo and an emu feature on Australia's Coat of Arms, and depictions of kangaroos are readily used in Australian currency, branding and logos. Some people feel it is inappropriate to eat such a significant national emblem. Others regard kangaroos as 'cute' and human friendly, associating them with characterisations in children's books and popular culture (most notably for many adult Australians, 'Skippy the bush kangaroo' from the 1968–1970s television series). Conservation and welfare are also of concern, with some people thinking that kangaroos should be protected along with many other native animals, for fear of them becoming endangered (Ampt and Owen 2008; Braddick 2002, 15; Cushing 2019, np; Probyn 2011, 36, 39, 49).

Conversely, kangaroo is more sustainable and environmentally low-impact than introduced species. Hard-hoofed livestock compact soils which in turn reduces the ground's capacity to absorb and retain water, thus exacerbating salinity and erosion, whereas kangaroos are soft-footed and move lightly over rangelands (Pascoe 2014, 24, 26, 43; Australian Bureau of Statistics (ABS) 2010). They emit less methane and greenhouse gases than sheep or cattle (Appleby 2010, 6, 38; Wilson and Edwards 2008, 119–128). As they are not farmed or dispatched in an abattoir, kangaroos have the benefit of being truly free-range, living a natural life until harvested by licensed hunters certified as employing 'humane' practices (Pople and Grigg 1999a; see also Ankeny and Bray 2018). 'Kangatarianism'—eschewing exogenous meats—has thus become an ethical meat-eating option (Cushing 2019, np). Relatively high in protein, leaner and lower in cholesterol than other red meats, kangaroo is also valued for its health benefits.

Although stigmatisation of kangaroo meat is lessening, studies show that practical barriers alienate some consumers. Having to cook kangaroo meat with care and serve many cuts rare and the relatively strong savoury flavour and gamey smell of kangaroo compared with lamb and beef are off-putting for some people (Waitt and Appleby 2014). Visibility and availability have been barriers (Ampt and Owen 2008, ix). Few mainstream cookbooks feature kangaroo recipes, and until relatively recently only selected supermarkets stocked kangaroo. Kangaroo steaks, fillets, mince, sausages and other value-added products are now increasingly obtainable in mainstream metropolitan supermarkets.

Indigenising Lamb

As Australia Day approaches each year, marketers promote iconic Australian foods such as Lamington cakes and 'damper' bread loaves; some products are emblazoned with the Southern Cross or Australian flag; and McDonald's offers an Aussie Burger, distinguished by the addition of bacon and beetroot. No particular food tradition has been universally adopted to mark the event in the way that turkey has with Thanksgiving in America or plum pudding at Christmas in Britain. Recognising that 'Australia Day was vacant—no brand "owned" it' (Effie Australia 2009), MLA promoted lamb in 2002 as 'Australia's Fare' and have produced long-form Australia Day–themed commercials each year since 2006, reaching millions of viewers on television and online (Centre for International Economics 2008, 18).[9]

Lamb might indeed seem an appropriate choice for Australia's national meat. It meets the criteria that historian Rachel Laudan applies to a national cuisine: it is familiar, readily available, eaten across the country and 'assumed to have a long continuous history, and to reflect and contribute to the national character' (2013, 324). Sheepmeat was the mainstay of the Australian diet for almost 150 years, and for much of Australia's settler colonial history, the country's wealth rode 'on the sheep's back' (Anonymous, 1924, 11)—primarily from the wool industry. Flocks were at first so precious that despite owning 1000 sheep in 1795, wool industry pioneer, John Macarthur, refused to kill any for meat, his table

[9] The annual campaign has been enormously successful and 'has become as iconic as lamb itself'; the 2016 advertisement was viewed over 5.5 million times online (MLA 2016).

generously supplied with kangaroo and wild duck (Onslow-Macarthur, 1914). By the 1840s there was 'mutton galore', mutton meat deemed by one commentator, 'merely the soil on which wool is grown' (Mundy 2006, 90, 176). Sheep brought wealth and power to the mighty 'squattocracy'—pastoralists and graziers—who, with a voracious appetite for land, assumed and appropriated great tracts of Aboriginal peoples' country with colonial certainty, defending their flocks and imperially sanctioned territorial rights. Many were absentee landholders who left armed stockmen to fend off traditional owners—often violently, sometimes cruelly—who posed a threat or showed resistance to the colonists' presence and activities (Mundy 2006, 79–82).

By 1891 there were 75 million sheep being run in Australia for the wool and meat trade, with relatively recent artificial freezing technologies enabling the export of meat to Britain (Roberts 1891). The value placed on sheep is immortalised in the 1895 bush-ballad still sung today, Waltzing Matilda (see Gregory n.d.), in which an opportunistic swagman (itinerant worker) chooses to drown himself in a billabong (waterhole) when caught by troopers with a 'jumbuck' (sheep) he had secreted in his 'tucker bag'. Had he killed a kangaroo there would have been no story.

'The Meat That Doesn't Discriminate'

Today, lamb has become a relatively prestigious meat. Prices have increased in recent years, mainly due to export demands, and overall consumption of red meat has dropped as chicken and pork have become cheaper options. Even so, Australians consumed 6.8 kg of sheepmeat per capita in 2019—five times the global average (MLA 2020, 5, 7; see also Whitnall and Pitts 2019).[10] Advertising has played a significant part in associating lamb with Australian-ness. 'Ethnographic research' conducted for marketing purposes revealed that lamb appealed to Australians' sense of sociability and was 'the only meat to genuinely cross Australia's ethnic diversity' (Buresti 2018, 5–6). While the MLA campaign strategy was for lamb to play a positive role of inclusivity in Australian culture (Buresti 2018, 5, 7), the light-hearted and often irreverent approach to Australian social

[10] Beef consumption was 25 kg, twice the global average.

identity portrayed in Australia Day Lamb and other MLA advertisements has often caused controversy: for insensitive cultural and religious stereotyping, ridiculing vegetarians and vegans and, more recently, trivialising the impact of colonisation. The 2017 advertisement blithely features Aboriginal Australians barbequing lamb on a beach while welcoming boatloads of colonisers and immigrants, all partying together once ashore.

The 'meat that doesn't discriminate' (MLA 2016) and the 'most uniting meat in the world' (Buresti 2018, 6–7) may be a culinary icon in Australia, but it has an awkward fit in terms of being iconically Australian. As an introduced food resource which is highly destructive to the environment, lamb is resoundingly un-Australian. Presenting lamb as the national celebratory meat emphasises the pervasive persistence of settler colonialism. To then claim lamb as unifying and non-discriminatory can be seen as an example of the structural invasion process and settler colonists' self-indigenisation. Lamb has helped settler colonists to 'eliminate', 'destroy' and 'replace' the indigenous land- and food-scapes (following Wolfe 2006, 387, 388; Rowse 2014, 299). But rather than being a symbol of settler colonialism, expropriation and oppression, it is used to promote the nation as a mature and advanced liberal democracy (following Youé 2018, 70). While not wanting to oversimplify the complexities of the structural framework of settler colonialism, lamb can be viewed as metaphor for the settler colonial paradigm outlined by Veracini, in which 'successful settler colonies "tame" a variety of wildernesses [and] effectively repress, co-opt, and extinguish indigenous alterities' (2011, 3). As outlined above, sheep played an integral role in the shaping of colonial and settler colonial Australia, in physical, socio-political and economic senses. Pastoral expansion generated wealth that helped enable independent government and processes that obscured and ruptured Aboriginal people's traditional lifeways, including their food production practices. Aboriginal people's health and morale 'plummeted' as their croplands were 'mown down' by introduced livestock; 'sheep, which walked ahead of their shepherds, helped eliminate evidence of [Aboriginal] agriculture and its domesticates' (Pascoe 2014, 18). This forced Aboriginal people to join or become dependent on the colonial system to survive (see van Reyk, 2021, 111–115). Separation from local knowledge, skills, traditions, lands and/or language profoundly disempowered (and continues to disempower) Aboriginal people—a recognised and typical outcome of structural settler colonialism (Wolfe 2006, 395).

Figures 9.1, 9.2 and 9.3 give an idea of the extent to which sheep and lamb production have impacted Aboriginal people's land and food growing areas.

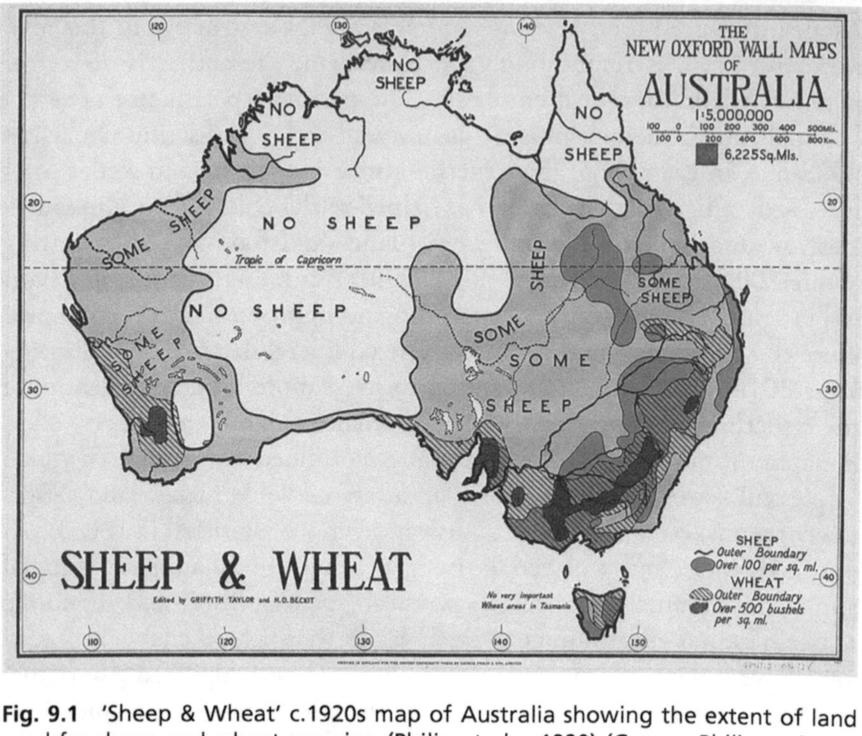

Fig. 9.1 'Sheep & Wheat' c.1920s map of Australia showing the extent of land used for sheep and wheat growing (Philip et al. c.1920) (George Philip and Son, and Taylor, Thomas Griffith, and Beckit, H.O. National Library Australia http://nla.gov.au/nla.obj-234331946. An equivalent 'Cattle and Minerals' map (National Library of Australia MAP G8961.J1 [192–?] https://nla.gov.au/nla.obj-255201355/view) adds to the picture. For more current distribution of sheep stocks, see MLA map 'Sheep numbers as at June 30 2019' (Accessed March 30, 2021). https://www.mla.com.au/globalassets/mla-corporate/prices%2D%2Dmarkets/documents/trends%2D%2Danalysis/fast-facts%2D%2Dmaps/2020/mla-sheep-numbers-map-2020-at-june-2019.pdf)

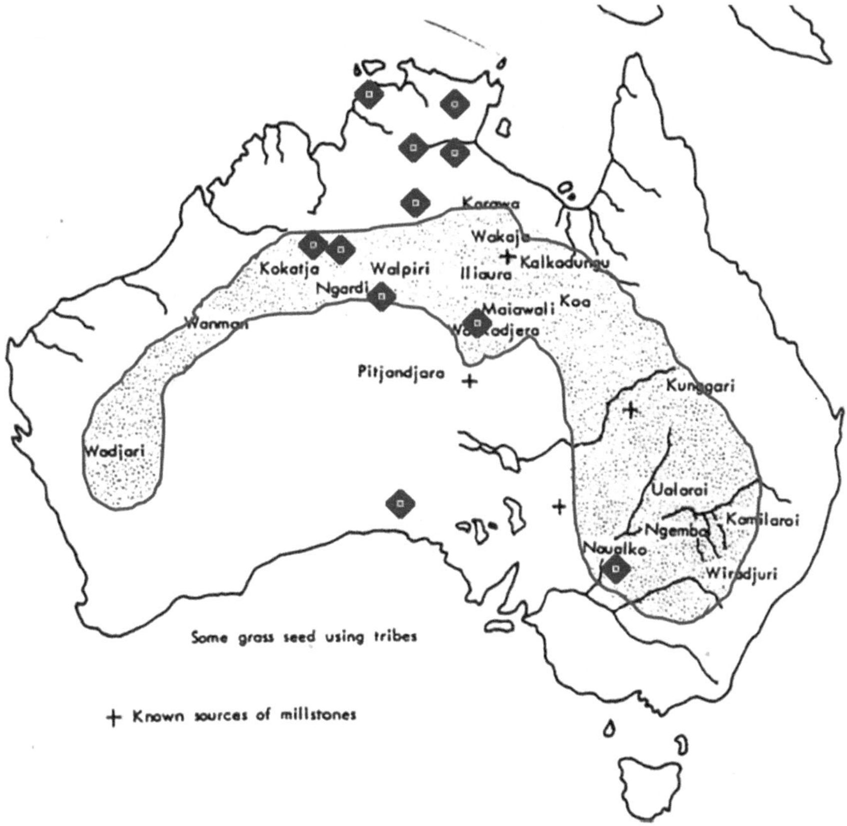

Fig. 9.2 'Tindale's Arc' (Chivers et al. 2015, 6, following Tindale, 1974) showing grassland areas traditionally used by Aboriginal people for grain. According to Chivers et al., other evidence suggests that grain processing was 'much more widespread than suggested in [Tindale's] map' (2015, 5, 6). Viewed in parallel with Fig. 9.1, we see the potential impact of sheep on Aboriginal people's traditional foodscape, in Australia's eastern and western regions.

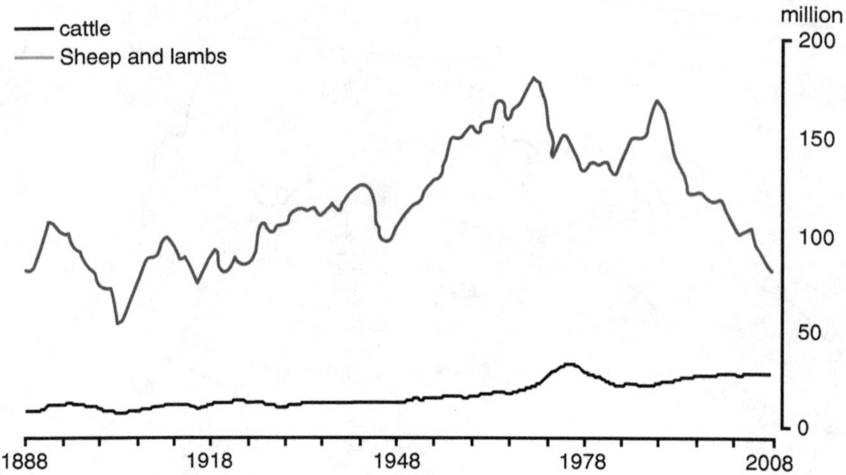

Fig. 9.3 Livestock grazing pressures 1888–2008 (ABS) 2010, n.p.). According to the ABS, 'agriculture is the most extensive form of land use in Australia [and] livestock grazing accounts for the largest area of land use in agriculture.' (2010, n.p)

Eating Otherness

The next stage in the settler colonial process, according to Veracini, is to 'effectively manage ethnic diversity' (2011, 3). Lamb conveniently found itself a place at the multicultural table, but its validity as an iconic Australian meat and its consumption as patriotic can be openly contested during the phase in which 'anti-colonial rhetoric express[es] a demand for indigenous sovereign independence and self-determination' (Veracini 2011, 5).

Settler colonial Australians have broadened their tastes and pride themselves for being open to the foods of others as a form of culinary liberalism, but native foods have only gained attention relatively recently. The lack of acceptance of and low regard for native foods compared to geographically and culturally exotic foods has been attributed to prejudice against Aboriginal people (Newton 2014, 244). In food scholar John Newton's view, 'Australian multiculinarity—accepting the food of "the others", eating our neighbours' food—has helped ease our way into what is generally a remarkable multicultural stability' (2016, xi). Applying the same theory to eating Indigenous foods, Newton postulates that 'surely, then, accepting the food of this land' will contribute towards 'culinary reconciliation' (2016, xii). However, readily accepting native foods into one's diet does not necessarily signify social inclusion and respect for Aboriginal people (Veracini 2011, 7–8; Craw 2012, 19–20).

Commodifying native foods for their difference and otherness leads to questions of appropriation rather than appreciation (hooks, 1992, 380; Craw 2008, 53; Craw 2012, 21). Following racial activist bell hooks' observations of white people's attraction to otherness (1992, 21–39), eating native foods as a conscious act may, in the consumer's mind, assuage the guilt associated with the nation's colonial past and deny accountability for Australia's wrongdoings towards Aboriginal people (Craw 2012, 19–20). Indigenous Studies scholar Charlotte Craw asserts that this renders native foods into 'fetish objects that promise an easy salve for colonial guilt without requiring actual change on the part of settlers' (Craw 2012, 21). Other scholars and chefs support the idea that native ingredients should be used like any other, without distinguishing them as native (Newton 2016, 110). While lauding them for their flavour or other merits, this denies any recognition for First Nations people and their age-old knowledge and practices. Few commercial manufacturers are owned or managed by Indigenous community members (Foley 2005, 28), although some producers work in consultation or collaboration with local communities (see Mann 2021, 136–140; Newton 2016, 177–181; van Reyk 2021, 236–240). It is generally agreed that unless Indigenous Australians benefit from all levels of production and supply, commercial adoption of native produce can only be classed as appropriation (Craw 2008, 52–53; 2012, 21). Aboriginal scholar Bruce Pascoe delivered a challenge to non-Indigenous producers and consumers in 2018: 'You can't eat our food if you can't swallow our history' (Allemann 2018, 4). This history includes the development of food systems that sustained Aboriginal people for over 60,000 years, the results of which are now targeted for commercial exploitation. The genetic value of Indigenous plants for food and agriculture is now recognised, and there are moves towards restricting access rights and legally protecting First Nations people's traditional culture and knowledge through intellectual property laws (Stoianoff and Roy 2015, 763).

Australia today is, unarguably, a settler colonial state that has been developing since 1788 to the detriment and exploitation of Aboriginal peoples and their culture. 'Culinary reconciliation' may taste good on the palate, but it is difficult to imagine that having a cuisine featuring or founded upon native produce might free settler colonial Australians from what Veracini deems an 'unsettling anxiety' (2011, 3). Given the gravity of subjectification and atrocities suffered by Aboriginal people since colonisation, the notion of culinary reconciliation seems to me grossly insulting to First Nations people. Echoing my 2011 self, eating native food on Australia Day seems in many ways patriotic and celebratory of Australia. It has also been suggested that January 26 be celebrated with 'a meal of native Australian foods shared between European and Aboriginal Australians … giving thanks [following Pascoe 2014, 156] to the Indigenous inhabitants for caring for the country, and [for] showing us the foods of the land' (Newton 2016, 228). Without responses from members of Indigenous communities, it is impossible to know how this well-intentioned but idealised display of appreciation and unity would be regarded in terms of reconciliation beyond the culinary. Unless the invitation to dine together is extended from Indigenous community members on their terms, gratitude from those who have gained privilege from the exploitation of others at this shared table may be as unpalatable as the Thanksgiving meal is to many American First Nations' people, especially while social and economic inequity remains and disregard for or denial of their history continues. Enjoyment and appreciation of Australian native foods may be a step forward in acknowledging Indigenous people and their culture, both traditional and contemporary, but it is ambitious (or arrogant and patronising) to imagine it will heal the hurt that has been inflicted upon First Nations communities, and in many cases, continues, or is exacerbated by new forms of exploitation.

Rethinking, Reframing, Reclaiming

Distancing themselves from the implications of Australia Day for First Nations people, MLA has rebadged its January national celebrations advertisement 'Australian Lamb'. The campaign continues to promote lamb as an emblem of social and cultural unity, which Australia Day is officially meant to engender and foster. When seen in the context of settler colonialism, lamb is highly compatible with 'invasion'—a symbol of imperial legacy and settler colonial objectives. Presenting lamb as the national celebratory meat supports a broader legacy of whitewashing in and of Australia's history and demonstrates the self-indigenised settler colonial view of Australia and Australian identity. Many would agree with leading chef and food writer, Stephanie Alexander, that 'one cannot imagine anything more Australian than Kangaroo' (Alexander 1996, 357). But while January 26 remains the national day of celebrating 'as one', knowing what the day represents to Australia's First Nations people who continue to fight for constitutional recognition, lamb as the quintessential settler colonial meat seems more appropriate than kangaroo, and I recognise now my idealised and supposedly 'woke' proposal from ten years ago as self-serving and tokenistic.

Native foods are slowly joining the array of foods on Australian tables predominantly as ingredients rather than dishes *per se* and rarely as they were traditionally prepared or consumed in their original contexts. Kangaroo might occasionally be served with warrigal greens and quandong relish or bush tomato infused barbeque sauce, but it is more likely a local substitute for other meats, doused in soy sauce for a stir-fry, sautéed in Mexican spices for burritos or sandwiched in a hamburger bun with the mandatory slices of beetroot. Native plants and seeds that would never naturally grow in 'cooee' (close geographical proximity) of each other are combined into bush spice mixes for any choice of use. Whether consumers choose them for their flavour, for health benefits as superfoods or in acknowledgement of their provenance as a self-conscious form of place-based pride or support for First Nations people, there are blurry lines between appreciation, appropriation and acculturation. Some may consider this 'culinary liberalism' as a sign of decolonisation by the dominant culture, where all foods and ethnicities blend together as one, on equal terms, satisfying settler colonial ideals of sharing and celebrating an ancient and ongoing Indigenous culture that settler colonial society has subsumed rather than erased. Writing about Australian culinary identity and cuisine, Greenwood says that 'as long as our culture fails to reflect the Aboriginal experience, so too will our cuisine' (2011, 68). The 'Aboriginal experience' is of course profoundly complex; it cannot be reduced to a single entity. It is both historical and very much in the present (see Foley 2005).

Native foods most certainly have a role to play in expressing Indigenous identity on the plate. In the hands of Indigenous cooks, they may embody the 'Aboriginal experience' as products of an ancient land and age-old cultures and of Indigenous people's knowledge and survival in a contemporary world. Archival and contemporary footage of Aboriginal people cooking fish, goanna or kangaroo whole and un-gutted in the traditional manner on campfires or in firepits can be readily found online and are shown in the media. Continuing these traditions is an important part of preserving Indigenous identity and heritage, but not all Indigenous people live off the land or cook in traditional ways on a day-to-day basis. To expect or assume that Indigenous people live on a seemingly crude 'bush' diet rather than mainstream foods serves to denigrate, historicise and stereotype Indigenous people and culture (Foley 2005, 27–29). Chefs, Indigenous and non-Indigenous, have contemporised native foods well beyond 'bush tukka', creating innovative dishes in hatted and starred restaurants. While colonisers have been criticised for cooking local ingredients using European techniques, today's chefs are incorporating native foods and flavours into restaurant classics with menus featuring confit muttonbird mousse, pumpkin and kunzea gnocchi, lemon myrtle panna cotta, bunya nut hummus and Geraldton wax grissini, while native botanicals are infused into spirits and give cocktails a local twist (Charcoal Lane, 2021; see also Foley 2005, 30; Craw 2012, 16, 19–20). These applications reflect the contemporary experience of Australia's First Nations people who are retaining, or for many, (re)discovering, their Indigenous heritage—embracing, honouring and celebrating their Indigeneity within and as part of this settler colonial, multicultural, polyglot society—through food. In time, one of their creations may become Australia's national dish, and truly represent Australia as an inclusive, integrated society, on the plate.

Bibliography

Abbott, Edward. 1864. *The English and Australian Cookery Book: Cookery for the Many and the Upper Ten Thousand.* Ludgate Hill, London: Sampson, Low, Son & Marston.

Aboriginal Land Council of New South Wales. 2021, January 15. Voice Proposal – Good Start But Falls Short of Uluru Aspirations. Accessed January 23, 2021. Voice Proposal – Good Start, But Falls Short of Uluru Aspirations | NSW Aboriginal Land Council (alc.org.au).

Act to Protect the Pastures and Live Stock of the Colony from the Depredations of Certain Noxious Animals 1880. (New South Wales). Accessed December 29, 2020. http://www.austlii.edu.au/au/legis/nsw/num_act/paspa1880n32363.pdf.

Alexander, Stephanie. 1996. *The Cook's Companion.* Ringwood, VIC: Viking, Penguin Books.

Allemann, Samantha. 2018, May. Indigenous Foods: Swallowing Our History. *Matters Journal*: np. Accessed November 20, 2021. https://www.samantha-allemann.com/food/swallowing-our-history-matters-journal/.

Ampt, Peter, and Owen, Kate. 2008. Consumer Attitudes to Kangaroo Meat Products. Rural Industries Research and Development Corporation. 2008, Section 4.2.3, 4.2.4, 5.5.

Anderson, Lara, and Merle Benbow. 2015. Cultural Indigestion in Multicultural Australia: Fear of "Foreign" Foods in Australian Media. *Gastronomica* 15 (1): 34–43. University of California Press.

Ankeny, Rachel, and Heather Bray. 2018. Ferals Or Food? Does Hunting Have a Role in Ethical Food Consumption in Australia? In *Wild Animals and Leisure: Rights and Wellbeing*, ed. N. Carr and J. Young, 210–224. Abingdon, Oxon, UK: Routledge.

Anon. About Town. *Daily Telegraph* (Sydney). Wednesday 30 July 1924, page 11. Accessed November 21, 2020. https://trove.nla.gov.au/newspaper/article/245211043?.

Appleby, Bryce. 2010. Skippy the "Green" Kangaroo: Identifying Resistances to Eating Kangaroo in the Home in Context of Climate Change. [BSc Hons] School of Earth & Environmental Sciences, University of Wollongong.

Australian Bureau of Statistics. 2010. Land Use. 4613.0 – Australia's Environment: Issues and Trends. Accessed December 29, 2020. https://www.abs.gov.au/AUSSTATS/abs@.nsf/Lookup/4613.0Chapter95Jan+2010.

Australian Human Rights Commission. 2020. "1938 Day of Mourning" and Online Links. Accessed January 16, 2021. https://calendar.humanrights.gov.au/1938-day-mourning.

Australian Institute of Aboriginal and Torres Strait Islander Studies (AIATSIS). n.d. We Hereby Make Protest. Accessed January 16, 2021. https://aiatsis.gov.au/explore/day-of-mourning.

Bannerman, Colin. 2011, November. Making Australian Food History. *Australian Humanities Review*: np. ANU ePRESS.

———. 2019. *The People's Cuisine: Origins of Australian Cookery*. Fremantle: Vivid Publishing.

Barnard, Marjorie. "Mask of Australia." *The Home* (March 1938): 25–27, 71. Sydney: Art in Australia.

Bennett, Lindsay. 2018, January 11. A Look Through 12 Years of MLA's Epic Summer Ads. AdNews, Yaffa Media. Accessed November 21, 2020. http://www.adnews.com.au/news/a-look-through-12-years-of-mla-s-epic-summer-ads.

Braddick, Lynda. 2002. Market Place Demand for Kangaroo Meat Consumption in Western Australia: A Sustainability Issue. Honours Thesis, Curtin University.

Buresti, Fabio. 2018. How Bringing People Together Delivered Record Profits for Lamb. Effie Awards Australia, The Monkeys Long Term Effects Entry. Accessed January 6, 2021 (Since Paywall Protected). https://www.effie.org/case_database/case/AU_2018_76.

Cavanaugh, Edward. 2012. History, Time and the Indigenist Critique. *Arena Journal* 37/38: 16–39.

Centre for International Economics. 2008, December. An Evaluation of Domestic and US Lamb Marketing. Report Prepared for Meat and Livestock Australia. Accessed January 16, 2021. https://www.mla.com.au/globalassets/mla-corporate/generic/about-mla/an-evaluation-of-lamb-on-farm-programs.pdf.

Chivers, Ian, Warrick, Richard, Bornman, Janet, and Evans, Chris. 2015. *Native Grasses Make New Products: A Current Review of Past Uses and Assessment of Potential*. Publication Number 15/056. Rural Industries Research and Development Corporation (RIRDC) Canberra. https://www.agrifutures.com.au/wp-content/uploads/publications/15-056.pdf

Commonwealth Franchise Act, 1908. Accessed February 20, 2021. https://www.legislation.gov.au/Details/C1902A00008.

Costantoura, Paul. 2020, January 30. The MLA Ad Confirms Australia Day Is the Latest Victim of Cancel Culture. Mumbrella (Online). Accessed

November 21, 2020. https://mumbrella.com.au/the-mla-lamb-ad-confirms-australia-day-is-the-latest-victim-of-cancel-culture-614913.

Craw, Charlotte. 2008. The Flavours of the Indigenous: Branding Native Food Products in Contemporary Australia. *Sites, Journal of Social Anthropology and Cultural Studies* 5 (1): 41–62.

———. 2012. Gustatory Redemption? Colonial Appetites, Historical Tales and the Contemporary Consumption of Australian Native Foods. *International Journal of Critical Indigenous Studies* 5 (2): 13–24.

Cushing, Nancy. 2019. To Eat Or Not to Eat Kangaroo: Bargaining Over Food Choice in the Anthropocene. *M/C Journal* 22 (2): np.

Dale, David. 2020, January 13. The Great Australian Bite: In Search of Our National Dish. *Good Food*. https://www.goodfood.com.au/eat-out/news/the-great-australian-bite-in-search-of-our-national-dish-20191218-h1kivx.

Effie Australia. 2009. Lamb on Australia Day – Defying the Law of Diminishing Returns One Chop at a Time. Meat and Livestock Australia, Effie Awards submission 2007–2009. Accessed November 21, 2020 (Since Paywall Protected). https://advertisingcouncil.org.au/effies/case-histories/lamb-on-australia-day-2007-2009-defying-the-law-of-diminishing-returns/.

Flowers, Rick, and Elaine Swan. 2012. Eating the Asian Other? Pedagogies of Food Multiculturalism in Australia. *PORTAL Journal of Multidisciplinary International Studies* 9 (2): np.

Foley, Wendy. 2005. Tradition and Change in Urban Indigenous Food Practices. *Postcolonial Studies* 8 (1): 25–44. https://doi.org/10.1080/13688790500134356.

Granville, Chandos, Thomas Baring, F.T. Buckland, C. Wentworth Dilke, and Thomas Fairbairn. 1860. Journal of the Society for Arts, Vol. 9, No. 419. *The Journal of the Society of Arts* 9 (419): 19–36. *JSTOR*. Accessed 13 December 2020. www.jstor.org/stable/41334090.

Greenwood, Helen. 2011. Migrants and Mavericks: Defining Australian Cuisine. In *Voracious*, ed. Paul McNally, 60–69. Prahran, VIC: Hardie Grant Books Australia.

Gregory, Mark. n.d. Waltzing Matilda. *Australian Folk Songs*. Accessed January 30, 2021. http://folkstream.com/094.html.

hooks, bell. 1992. *Black Looks: Race and Representation*. Boston: South End Press.

Instone, Lesley. 2005. Eating the Country. *Journal of Australian Studies* 29 (86): 135–141. https://doi.org/10.1080/14443050509388039.

Lancelott, Francis. 1852. *Australia as It Is: Its Settlements, Farms and Goldfields*, 1. London: Colburn & Co. Google Books Electronic Book, Digitised

September 26, 2007. Accessed November 11, 2020. https://books.google.com.au/books?id=FsMNAAAAQAAJ&vq.

Laudan, Rachel. 2013. *Cuisine and Empire, Cooking in World History*. California: University of California Press.

Levi-Strauss, Claude. 1990. *The Origin of Table Manners* [English Translation of *L'Origine des manières de table*, by John Weightman and Doreen Weightman, 1978]. Chicago: University of Chicago Press.

Macarthur, Elizabeth. Letters Dated 23 August 1794; 1 September 1795 in Onslow, Sibella Macarthur (editor). 1914. *Some Early Records of the Macarthurs of Camden*, Angus & Robertson Sydney. Project of Gutenberg Australia e-Book, Updated 2014. Accessed April 11, 2021. https://gutenberg.net.au/ebooks13/1302011h.html.

Maclurcan, Hannah. 1905. *Mrs Maclurcan's Cookery Book: A Collection of Practical Recipes, Specially Suitable for Australia*. Sydney: George Robertson and Company.

Mann, Alana. 2021. *Food in a Changing Climate*. Bingley: Emerald Publishing Limited.

McKay, David. 2017, June 19. Uluru Statement: A Quick Guide. Parliament of Australia. Accessed January 16, 2021. https://www.aph.gov.au/About_Parliament/Parliamentary_Departments/Parliamentary_Library/pubs/rp/rp1617/Quick_Guides/UluruStatement.

McKenna, Mark. 2012, August 8. History and Australia: A Foundational Past?" *The Annual History Lecture, History Council of New South Wales*. Accessed November 11, 2020. https://historycouncilnsw.org.au/wp-content/uploads/2013/01/2012-AHL-McKenna.pdf.

Mundy, Godfrey Charles. 2006. *Our Antipodes*, [1852]. Abbreviated and with an Introduction by D.W.A. Baker. Canberra: Pandanus Books.

Muskett, Philip. 1893. *The Art of Living in Australia*. Sydney: Eyre and Spottiswoode.

Newling, Jacqui. 2011a. Crown Lamb. Proceedings of the 18th Symposium of Australian Gastronomy, Canberra. http://www.gastronomers.net/wp-content/uploads/2011/05/AGSYMP18-Proceedings-HiRes.pdf.

———. 2011b. Dining with Strangeness: European Foodways on the Eora Frontier. *Journal of Australian Colonial History* 13: 27–48.

Newton, John. 2014. Terra Nullius, Culina Nullius. Doctoral Thesis, University of Technology, Sydney (Unpublished).

———. 2016. *The Oldest Foods on Earth*. Sydney: NewSouth Publishing.

———. 2018. *The Getting of Garlic*. Sydney: NewSouth Publishing.

O'Brien, Charmaine. 2016. *The Colonial Kitchen: Australia 1788–1900*. Lanham, MD: Rowman & Littlefield.

Pascoe, Bruce. 2014. *Dark Emu: Black Seeds: Agriculture Or Accident?* Broome, Western Australia: Magabala Books.

———. 2016. The Place and the Hearth. *Jonathan Jones: Barrangal Dyara (Skin and Bones)*, Jones, Jonathan, and O'Callaghan, Genevieve, 87–91. Lilyfield, New South Wales: Kaldor Public Art Projects.

Pearson, Margaret. 1894. *Australian Cookery Book for the People*. Melbourne: NP.

Pople, Tony, and Grigg, Gordon. 1999a. Population Monitoring, Quota Setting and Annual Harvests. *Commercial Harvesting of Kangaroos in Australia*. Chapter 4. Canberra: Department of Agriculture, Water and the Environment. Accessed December 26, 2020. https://www.environment.gov.au/node/16675.

———. 1999b. Kangaroos as Pests, Kangaroos as Resource. *Commercial Harvesting of Kangaroos in Australia*. Chapter 7. Canberra: Department of Agriculture, Water and the Environment. Accessed December 26, 2020. https://www.environment.gov.au/node/16678.

Probyn, Elspeth. 2011. *Carnal Appetites* Elspeth, "Eating Roo: of things that become food." *New Formations*, Number 74, Lawrence and Wishart: 33–45.

Rawson, Mrs Lance (Wilhelmina). 1895. *The Antipodean Cookery Book and Kitchen Companion* (Facsimile edition, 1992). Kenthurst, New South Wales: Kangaroo Press Pty Ltd.

Roberts, D. 1891. *Meat Export: Overstocked Runs: The Squatters' Threatened Ruin: The Only Substantial Relief: A Practical Economical System: Method of Working: Initiation of the System*. Sydney. Digitised Version https://nla.gov.au/nla.obj-475652103.

Rowse, Tim. 2014. Indigenous Heterogeneity. *Australian Historical Studies* 45 (3): 297–310. https://doi.org/10.1080/1031461X.2014.946523.

Rutledge, Jean. 1904. *The Goulburn Cookery Book*. NP.

———. 1937. *The New Goulburn Cookery Book*. NP.

Santich, Barbara. 1996. *Looking for Flavour*. Adelaide: Wakefield Press.

———. 2011. Nineteenth-Century Experimentation and the Role of Indigenous Foods in Australian Food Culture. *Australian Humanities Review* 51: 65–78.

———. 2012. *Bold Palates*. Adelaide: Wakefield Press.

Singley, Blake. 2012. 'Hardly Anything Fit for Man to Eat': Food and Colonialism in Australia. *History Australia* 9 (3): 27–42.

Stoianoff, Natalie, and Alpana Roy. 2015. Indigenous Knowledge and Culture in Australia – The Case for Sui Generis Legislation. *Monash University Law Review* 41 (3): 745–785.

Summers, John. 2000. The Parliament of the Commonwealth of Australia and Indigenous Peoples 1901–1967. Accessed January 16, 2021. https://www.aph.gov.au/About_Parliament/Parliamentary_Departments/Parliamentary_Library/pubs/rp/rp0001/01RP10.

Symons, Michael. 2007. *One Continuous Picnic*. Melbourne: Melbourne University Publishing.

Tibbertsma, Anita. 2017, January 27. How Brands Get Behind Australia Day. *BigDatr*. Accessed November 21, 2020. https://bigdatr.com/blog/how-brands-get-behind-australia-day.

Tindale, Norman. 1974. *Aboriginal Tribes of Australia – Their Terrain, Environmental Controls, Distribution, Limits, and Proper Names*. Canberra: Australian National University Press.

Trigger, Rebecca. 2017, September 5. Meat and Livestock Australia Criticised for Advert of Vegetarian Hindu God Ganesha. *ABC News*. Accessed January 16, 2021. https://www.abc.net.au/news/2017-09-05/lamb-ad-under-fire-for-portrayal-of-vegetarian-god-ganesha/8875048.

Van Reyk, Paul. 2021. *True to the Land: A History of Food in Australia*. London: Reaktion Books.

Veracini, Lorenzo. 2011. Introducing Settler Colonial Studies. *Settler Colonial Studies: A Global Phenomenon. Settler Colonial Studies* 1 (1): 1–12. https://doi.org/10.1080/2201473X.2011.10648799.

Waitt, G., and Bryce Appleby. 2014. 'It smells disgusting': Plating Up Kangaroo for a Changing Climate. *Continuum* 28 (1): 88–100. https://doi.org/10.1080/10304312.2013.854863.

Watkie, A. H. (ed.). 1949. *The Housewives' Calendar of Puddings*, The South Australian Country Women's Association (Incorporated), Mount Gambier.

Whitnall, Tim, and Pitts, Nathan. 2019. Meat Consumption: Analysis of Global Meat Consumption Trends. Department of Agriculture, Water and the Environment. Last Reviewed 21 October 2020. Accessed December 29, 2020. https://www.agriculture.gov.au/abares/research-topics/agricultural-commodities/mar-2019/meat-consumption.

Wilson, George, and Melanie Edwards. 2008. Native Wildlife on Rangelands to Minimise Methane and Produce Lower Emission Meat: Kangaroos Versus

Livestock. *Conservation Letters* 1: 119–128. Australian Wildlife Services, Canberra, 2008. Accessed December 30, 2020. https://ro.uow.edu.au/cgi/viewcontent.cgi?article=1102&context=thsci.

Wolfe, Patrick. 2006. Settler Colonialism and the Elimination of the Native. *Journal of Genocide Research* 8 (4): 387–409. https://doi.org/10.1080/14623520601056240.

Youé, Chris. 2018. Settler Colonialism Or Colonies with Settlers? *Canadian Journal of African Studies / Revue canadienne des études africaines* 52 (1): 69–85. https://doi.org/10.1080/00083968.2018.1429868.

Websites

"Sheep & Wheat" [Map]. George Philip & Son, and Taylor, Griffith and Beckit, H.O. (eds). London: Oxford University Press [c192–?]. National Library of Australia, MAP G8961.J1 [192–?]. Digital Access Via TROVE, Accessed March 30, 2021. https://nla.gov.au/nla.obj-234331946/view.

Charcoal Lane [Restaurant]. Accessed January 30, 2021. https://www.charcoal-lane.com.au/menu.

Macro Meats. Accessed December 20, 2020. http://macromeats.com/sustainability/sustainability/.

Meat and Livestock Australia (MLA). 2016, April 6. MLA's 2016 Australia Day Lamb Advertising Campaign Most Successful on Record. Accessed January 23, 2021. https://www.mla.com.au/news-and-events/industry-news/archived/2016/mlas-2016-australia-day-lamb-advertising-campaign-most-successful-on-record/#.

———. State of the Industry Report, 2020. Accessed December 29, 2020. https://www.mla.com.au/globalassets/mla-corporate/prices--markets/documents/trends--analysis/soti-report/mla-state-of-industry-report-2020.pdf.

———. Spring Lamb 2016 "You never lamb alone" Advertisement Viewed January 16, 2021. https://www.youtube.com/watch?v=DmIkgo_5yO0.

———. Australia Day Lamb Ad 2017. Viewed January 16, 2021. https://youtu.be/fBTWc4i_Fhw.

———. Sheep Numbers as at June 30 2019. Natural Resource Management Region [Map]. Accessed March 30, 2021. https://www.mla.com.au/globalassets/mla-corporate/prices--markets/documents/trends--analysis/fast-facts--maps/2020/mla-sheep-numbers-map-2020-at-june-2019.pdf.

Part III

After Decolonisation?

10

"A Manly Amount of Wreckage": South African Food Culture and Settler Belonging in Ivan Vladislavić's *Double Negative*

Nitzan Tal

Ivan Vladislavić wrote his 2010 novel, *Double Negative*, in response to iconic South African photographer David Goldblatt's work. Like the photographer, Vladislavić uses short "snapshots" to alert his readers to the "uncanny phenomena that explode from the social fault lines" of Johannesburg (Flanery 2013). Eating habits and foodways, it turns out, are central among these phenomena. Thus, *Double Negative* reinscribes the Gordian tie between identity and foodways as it describes the maturation of Neville Lister, a white South African of British descent, during apartheid and in the immediate aftermath of its dissolution.

By preparing and consuming food in different manners throughout the novel, Neville embodies and exposes a tension inherent to South African white settler-colonial identity. On one end of this tension or paradox sprawls the "manly amount of wreckage" created by consuming jumbo prawns or a braai (Afrikaner barbecue); on the other end sit the "individual ramekins" in which Neville's family serve their soufflés.

N. Tal (✉)
Cornell University, Ithaca, NY, USA
e-mail: nt354@cornell.edu

R. Ranta et al. (eds.), *'Going Native?'*, Food and Identity in a Globalising World,
https://doi.org/10.1007/978-3-030-96268-5_10

Vladislavić, a prodigious creator and curator of South African cultural production, thus collates moments of culinary meaning-making which reveal a tension between self-indigenising and self-Europeanising settlerhood. This tension occurs under, and results from, an epistemic regime which perceives being "African" and being "worldly" as antithetical.

To lay the groundwork for this argument, a quick overview of how settlers arrived in what is now South Africa—and how those settlers fed themselves—is in order. Though the Cape was a strategic port for Portuguese ships starting in the fifteenth century, it wasn't until 1652 that a significant European permanent settlement occurred. Motivated at first only to secure a base camp for operations, the Dutch East India Company nevertheless began settling Dutch farmers on the land to grow the supplies needed for its ships. Displacing indigenous populations which resisted attempts at enslavement, these farmers imported large quantities of enslaved peoples from East Africa, India, Indonesia, and elsewhere (Ross 1983). Following troubles in Europe, the British invaded the Cape in 1795 and won control of it by 1805. The gradual slide from administrative trade port to settler community repeated itself as Britain settled increasing numbers of tradespeople in and around the Cape.[1] The Dutch-descendent Afrikaners migrated inland, taking with them the species that were imported as stock reserves for the shipping station (Swart 2010). They also picked up indigenous foodways, such as the curing of spiced meat into *biltong* or *droe wors* and grilling fresh meat on an open fire, or *braai*. These became the pillars of Afrikaner cuisine, which also incorporated spices traded in the Cape from the Dutch colonies in today's Indonesia and Malaysia, as well as from the British colonies, especially India.

[1] Though the point of the colony was the production of supplies as well as a port for maintenance and administration, many ingredients of the settlers' diets continued being imported. Among these, a British agriculturalist writing in 1905 counted "(1) Dairy produce and eggs, (2) fresh meat and poultry, (3) fodder of all sorts, (4) salted fish and smoked meat, (5) vegetables and fruit. Besides these, but not included in this paper, we have preserved meat, vegetables and fish; dried fruits, not being currants, raisins, or figs; tinned and bottled fruits, and butterine, all of which are not, strictly speaking, producible in the Colony, although they take the place in a great measure of goods produced here". To these he added the imported foods which "nature prevents our growing, such as rice, sugar, coffee and spice" (Nobbs 1905).

The stakes of both Dutch and British colonisation rose significantly with the discovery of diamonds and then gold in the late nineteenth century. Rapid and involuntary urbanisation followed, as South Africa's native men were forced off the land and into grim miner compounds, motivating a move away from millet- and sorghum-based diets and towards today's corn-based staples. In their absence, home farms were hard-pressed to produce more than needed for sustenance, making way for settler-owned commercial agriculture. The segregation, labour exploitation, and oppression of the non-white population established in British colonial times continued in force after the Afrikaner National Party rose to power in 1948 and established its system of apartheid. Prepared by black and coloured domestic workers, white cuisine reflected the group's economic privilege and better (though still limited) access to imported goods. However, the separationist ideal of apartheid was never achievable in the culinary realm: a popular street food of hollowed-out bread filled with mutton curry, for example, is said to have originated in the forbidding of Indian grocers from selling hot meals to black migrant workers, prompting them to disguise curry in a permissible loaf of bread (Karon 2018). Rich with the resonances of conquest, enslavement, and oppression, South African cuisine serves as ready material for Vladislavić, one of the country's most astute literary figures.

A Manly Amount of Wreckage: Masculinity, Territory, and Braai

Double Negative's attention to food and dining begins with that all-important ritual of South African whiteness, the barbecue, or braai. The Anglo Listers are joined by their newly arrived neighbours, the Afrikaner van Huyssteen family. Virility, masculinity, and South Africanness are measured by the braai, in Neville's memory, as he witnesses Mr van Huyssteen squeezing sausages on the grill and then sliding his hands between his wife's thighs. Neville's recollection betrays his teenage self's need to excuse his own family's Anglo timidity in comparison: "when it came to outdoor living we were not in the same league, but we had the patio and the pool, and my dad could char a lamb chop as well as the next man" (25). Already the novel intimates the layered colonial identities which comprise South African society: the tension between Afrikaner and British-descendent people, carrying an echo of the two "Anglo-Boer Wars" of the late nineteenth century.

Reading nineteenth-century British texts on South African Boers, or Apartheid-era Afrikaner depictions of frail, un-African "Anglos", one is tempted to conclude that "for the first third of the twentieth century the dominant form of racial antipathy in South Africa was between the two white ethnic groups" (Dubow 1989, 4). This is of course blatantly untrue; yet the focus and fervour of this "racial antipathy" played a decisive role in the self-indigenisation of the white minority in South Africa (see Veracini 2015; Du Toit 2014). The braai hosted in honour of the van Huyssteens serves as the memory propelling Neville's story—a vanishing point from which the rest of the novel emanates; the studied casualness which the scene describes is proof of the complex move from the historical animosity to the shared white identity demanded by apartheid's racial classification.

South African poet and scholar Marlene van Niekerk claims that Afrikaner nationalist pride utilises food culture, including the braai, as part of an ethnic preparedness for an envisaged struggle for self-preservation (2011, 63). A performance of identity as much as a meal, the braai carries expectations of male camaraderie and excess: "the big eat and showing off the big eat are inseparable" (van Niekerk 2011, 65, 69; see also Osseo-Asare 2005, 67). The braai, adapted from native grilling practices, nevertheless preserves connotations of the northward expansion of the Dutch descendants (*voortrekkers*) and belongs to other South African groups—like the British descendants—by association. It is not only the van Huyssteen's ribald leisure but also the pioneer valour attached to the "outdoor living" of the *voortrekkers* which Neville's family lacks (Vladislavić 25).

It is easy to read the braai's significance in its opulence, in the casually luxurious heaped plates: Neville's mother had put some food aside for him in a pointless gesture, since "there was so much left over" (Vladislavić 26). White privilege under apartheid meant white people could enjoy lavish meals while others hungered; that is it. This is, of course, part of the picture, but we would be missing the intricate construction of settler identity which supports this economic structure if we stopped there. A second food scene in the same part of the novel, in which characters convene around piles of the large shrimp that breed in nearby stretches of the Indian Ocean, suggests that the leftovers are as important as the meal consumed: that it is the remains which make some forms of eating performative of local settler identity.

In this second scene, Neville sits with famed photographer Auerbach—his would-be mentor—in a fish restaurant, bemoaning his order as he watches his compatriots devour their messy heaps of prawns. "I should have had prawns too," he thinks, "it would have given me reason to splash butter and lemon juice, to suck at my teeth and burn the hell out of my mouth and leave a manly amount of wreckage on the plate" (74). The braai and prawn scenes are connected through the shared sense of a gastronomic male camaraderie that is intimately tied to the creation of waste and in which Neville fails to participate. The braai's "heaped" plates and "overburdened" air become debris which the text enumerates: "paper plates full of bones and peels, crumpled serviettes and balls of foil, bloody juices… thick chops and coils of wors soldered to the stainless steel with grease" (29–30). It is not the eating but the remains which do the political work here, which seem to produce both the food's masculinity and its relationship to power. Not only does the cook "commandeer" the barbecue tongs, explicating the braaing act as a seizure of control (27); the excess created attaches itself immovably to the grill ("soldered"), the waste spills over its sides and takes over the scene, occupying the space. Though in his home setting our narrator connects this excess to the economic and political structure of apartheid, he nevertheless yearns to participate in it when it might win him a professional place at the grown-up table of the restaurant.

It follows that certain kinds of food, in certain settings, are cemented as masculine, and they are so gendered at least in part on account of the remains they leave behind, that "manly amount of wreckage". In the braai as well as prawn scenes, the novel casts these forms of eating as nonchalant and aggressive at the same time. These foods, and their consumption, are decidedly *local*: they are tied to locally sourced ingredients, bear endemic names, and as discussed are attached to a settler narrative of

South African history (Vladislavić 73). Their locality is further cemented, as I have shown, because the "wreckage" left behind takes up space—the eaters occupy the land even after the feasting is over.

Messy, masculine eating is a staple in many settler-colonial contexts. Cooking meat on an open fire seems to cater to settlers' needs of managing and occupying space, both during the process of colonisation and after its completion. Michael Symons argues the central role of meat in Australia's allure for nineteenth-century European settlers (2007, 28–41). In Argentina, Uruguay, and Brazil, cattle imported during the early sixteenth century have become a vector of national identity as well as a central object of export and sustenance. The *gauchos* roasting beef over an *asado* are considered national symbols of masculine frontier life and of idealised paternalism (Ehrick 2006; Slatta 1983; and Zycherman 2010). In the USA, meat consumption is closely tied to Manifest Destiny ideology (Willard 2002, 108). "[R]ugged American cowboys [were imagined as] staunch individualists, both territorialising land with their grazing animals and creating sustenance to conquer and take the West" (Scott 2010, 7). Nir Avieli's Israeli "Grilled Nationalism" posits a strong link between settler identity and the practice of an independence-day barbeque (2013).

Vladislavić's braai scene evokes these histories even as it describes ties of affect and materiality which are staunchly local. The yard and barbeque are idyllic, sun warmed slasto tile balancing the cooling evening, the pool water shifting "like a well-fed animal" besides the overladen plates (26). Yet, an ominous undertone runs through Vladislavić's text, an aggression emplaced within the idyll of the suburban environment. As the new neighbour Mr van Huyssteen tells of his childhood escapades torturing his parents' black servants, Neville can no longer enjoy the evening:

> The shift was imperceptible, as if someone had put on a record in the background, turned down low, and by the time you became aware of it your mood had already altered. An odourless poison leaked out of him. (28)

The same can be said for the text, which effectively communicates the unease belying the feast of grilled meats. Explaining why his father chose parquet floors for the house—perhaps the most bourgeois of internal debates—young Neville recalls his father thinks carpets turn any room into "a padded cell" (26). The neighbour whose children lie on the same parquet floor is, the reader knows, in the business of jail cells (28). And is not the "odourless poison" leaking out of Mr Huyssteen the same chlorine which the pool, a "well fed animal", breathed out? The laden paper plates become "debris… bloody juices" (29), and Mr Huyssteen's new car suddenly seems "an enormous piece of evidence", implying, of course, a crime (30). They had all eaten themselves sick, the riled-up Neville thinks, on "the fat of the land"—echoing the tainted comfort of Egypt's opulence as promised by Pharaoh (Genesis 45:18). Throughout the pages of the braai scene, this double character of comfort and latent violence emanates from the grill and its surroundings.

Neville's recollected discomfort at the braiding-together of grilling and violence has, besides the general social history of the braai, more immediate apartheid-era circumstances. The interviews of the Truth and Reconciliation Commission, conducted about a decade after Vladislavić's fictional scene takes place, revealed a perverse—but perhaps not altogether surprising—affinity between the ritual of the braai and the torture and execution of antiapartheid activists. Judge Andrew Wilson, who headed the amnesty committee which interviewed apartheid's perpetrators, remarked that he had "heard several cases where when people are being questioned the police who were doing the questioning [were] enjoying a braai" (TRC Hearing October 21 1996). In fact, the braai was so frequently a part of state-sanctioned violence that the terms became almost synonymous, one kind of burned flesh occasioning another.[2] As Alan Feldman puts it, braaing—always part of a "political culture of white male dominance"—integrated "consumption, commensality and violence" under apartheid, normalising scenes of torture while conversely

[2] Such were the cases of the "Pebco Three", whose torturers enjoyed a braai while they were "interrogating" them and whose bodies were subsequently burned and thrown in a river; of an activist known in the transcripts as T. Mvudle, who was branded with a meat fork used for braaing (TRC Hearings 11–14 November 1996); and of Charity Kondile, who testified how notorious murderer Dirk Coetzee "braaied" her son (qtd in Krog 2000, 307). Journalist and author Antjie Krog, who covered the TRC proceedings, recalls that a fellow writer complained that he does not want to use such language that mixes "breakfast and blood" (Krog 45).

refracting the penetration of terror into the culinary quotidian (Feldman 2002, 241–245).

Vladislavić's protagonist does not allude to the gruesome coupling of braai and torture that the text seems to hail, but he does describe the stakes of his own family's comfort in terms of the dinner table. "[J]ust imagine", Neville shouts at the neighbour, "that you've worked all your life down a bloody goldmine and you still can't afford to put food on the table" (30). For his infringement of the laws of hospitality and decorum—laws which are essential to the braai tradition—he receives a slap from his father, the first and last the father ever doles out. The slap across the face "knock[s] the world back into order" for Neville (31). This "order" is both familial ("The family motto had always been: 'Don't rock the boat'", 33) and political—white men should get to gorge peacefully; white boys should enjoy it. An apology is extended, hands are shaken. The Social, taking shape through the constellation of beer, braai, racial supremacy, and cordial hospitality at any cost, has collected its entry price.

There is enough at stake in the way food and its rituals are treated in *Double Negative* that the text itself comments on the theme. When in the last part of the novel a young journalist visits the grown Neville's home, they leaf through his mother's cookbook, and he explains that "[i]t's a bit of social anthropology too. The eating habits of the white middle class … under apartheid" (170, ellipsis in original). "Under apartheid" becomes an ironic term for Neville's cohort of white liberals, perhaps because, like the cookbook, it renders their lives an object for analysis. Neville's college history teacher had made a career out of this ironic distance, publishing "five books with 'under apartheid' in their titles" (143). The reflexive self-making of this "white middle class", in other words, engulfs the reader, who is included in the cycle of an almost banal political understanding of identity and food. Despite the novel's suggestion that these themes are too overt to be discussed further, I think their explication is worthwhile—particularly as the text formulates a tension vital to our understanding of settler existence and politics.

An intrinsic component of identity, eating is revealed to be a modality by which the hegemonic settler-colonial group adopts its paradoxical identity: self-styled Europeanness on the one hand and self-styled nativeness on the other (Hever 2007, 206; Zreik 2016, 358). The colonial "civilising" mission was predicated on the colonising population being better suited to governance of the land than its native. This advantage hinges variably on Europeanness, Christendom, whiteness, or other formulations of identity. However, this supremacy is challenged by the defining temporality of settler colonialism: a structure and not an event, wherein settlers "come to stay" (Wolfe 2006: 388). Settler identity, then, depends—in its dominant articulation—on an unresolved tension between European-esque and native-esque valences. The braai and the *cafetière*, which I will get to soon, stand as the material markers of these two identity functions in the novel, ultimately suggesting that genres of material comfort are endemic to being- and becoming-sovereign in the South African context.

It is appropriate, then, that South Africa's transition to democracy in the mid-1990s is marked by yet another braai scene. After ten years away, Neville returns to Johannesburg following the democratic elections. As a photographer, Neville sees behind the scenes of the manufacturing of this "new South Africa". He is moved by the images he helps create:

> I remember shooting stills on one of those rainbow nation commercials where a cheerful circle of friends, representing all the major population groups, gathered around the braai to drink beer and braai chops (but not yet to hold hands). These nation-building epics brought a lump to my throat, even if the easy companionship among the cast did not extend to the crew. When I left the studio and went back into the street, the present felt like the past. (112)

The transition years are "parenthetical", says Neville, "the old versions of things trail behind the new ones in brackets"—the rainbow nation may braai together, but only in a frozen image, and without touching. Temporally, narrating Neville is in the future, foreshadowing for the reader that the age of handholding will yet come. Neville's photography mirrors the novel's temporal elasticity: the image he produces creates a future, but Neville remains in the present, which feels like the past. The image of fellowship lures its creator, generating a distance between expectation and experience and thus, disappointment. I refer here to Andrew van der Vlies' claim that disappointment characterises life in the post-apartheid nation, as members find themselves "trapped in an imperfect present that is not as the future was imagined" (van der Vlies 2017, viii). For the reader, this scene mirrors the first braai, the one which almost made Neville a dissenter and which instead clarified his obligations to the political order. But, importantly, *ceci n'est pas un braai*. The multiethnic cheer is staged, and the chops are for show, not sustenance. Compared with the sensory abundance of the earlier scene, its "thick chops" and "overburdened air", this scene of "shooting stills" seems sterile (29; 27; 112).

The premises underlying Vladislavić's focus on material and culinary culture are that (a) the material world is a tangible expression of social relationships and (b) our relationship to our material surrounding is active and creative rather than passive—and this is true for fences and cutting boards, fruit bowls, and lamb chops (Beaudry et al. 1996, 272). Food, and the materials and rituals around it, are a constant reciprocal opening between us and the world. In the words of South African philosopher Marthinus Versfeld, "the material landscape gets into us via the kitchen and its pots" (Versfeld 1985, 27). The settler society of which Neville is a member creates food to create itself. To re-create itself, it stages an image of food, of a braai, in which no eating is done: this, the reader senses, will always be insufficient. The staged commensality does not hold the same effect on social relationship that mess-mates experience, nor can it eradicate the connotations of exclusionary and at times violent practices associated with the braai.

Before discussing what, if anything, may yet be learned from the reproduction of eating—a fitting description for Vladislavić's novel as well as the stilted braai scene described above—I must attend to a second kind of cuisine *Double Negative* explores. Equally important as masculinist, wreckage-producing feasts which communicate locality and "natural" ease are depictions of diminutive, elegant foodstuffs, those which tie the social group to an imagined European urbanity. Their stylistic counterpart in the novel is idioms of polite insistence. Both are encapsulated in the phrase Neville uses to narrate his dinner preparations, in which he and his wife serve salmon soufflé "in individual ramekins, if you don't mind" (Vladislavić 215).

Individual Ramekins, If You Don't Mind

The braai and prawn scenes were indicative of a specific, masculinist kind of material comfort which the novel tells us is an integral part of white South Africanness. But the novel describes another kind of sustenance, nurturing a complementary part of this bourgeois culture. If the braai, beer, and piri-piri sauce locate their consumers as local and at ease—complementing perfectly "the sated murmur of conversation, the outstretched legs and tilted head" that connote an informal and familiar locality—there exists a grammar of flavours, practices, and bodily gestures which do the obverse, locating their purveyors as cosmopolitan, refined, and fastidious (26).

Take Neville's wife Leora's preparations of fennel, baby salad, and salmon soufflé:

> Leora went back to chopping fennel on the butcher's block. It was Friday evening. The aromatic essence of her famous salmon soufflé—in individual ramekins, if you don't mind—came from the eye-level oven; a salad cut down cruelly in its youth, baby carrots, bean sprouts, young spinach leaves, lay in a bamboo bowl. While she mixed the dressing, I opened some wine (it was a compensatory Springfield Life from Stone, nursed to maturity in the rocky soils of the Robertson valley). (Vladislavić 215)

The self-aware description denotes upper-class consumption. It harkens towards France specifically and Europe in general, and it demands multiple adjectives and parenthetical enumerations. It is, at least in perception, a white affair. Contrast this scene with the shopping list Neville's mother finds crumpled in the supermarket cart—"mealie meal, pilchards, sticky tape, Doom" (maize flour, small fish, adhesive, and insecticide—p. 107). This second culinary world, hinted at in ruthlessly condensed list form and aware of the kitchen's less-savoury inhabitants, will have international readers of the novel reaching for a search engine. It is racialised non-white; it is neither Neville's shopping list nor his mother's. But it is home, to the extent that it merits collecting from its forgotten place in the cart and sending to Neville, the son living (at that point) in London. To explain the gap between what speaks of home (mealie meal) and what cooks at home (soufflé), we need to view Neville's artistic initiation as a cultural-demographic initiation into a certain settler class.

This induction into other-than-local aspiration of white middle-classness is intimated before Neville steps foot in Saul Auerbach's house. Older, narrator-Neville describes the setting in the luscious terms of far-away cuisines: "There was something Mediterranean about the dappled pergolas, the walls as creamy as feta, the succulent shadows of fig leaves and thick-tongued aloes cast by the late-summer sun" (43–44). The house seems edible with its tongues and feta, its succulence, and its figs. While he recognises it as foreign, it is clearly to Neville's taste. Like the braai which is the centre of the Lister's backyard, it radiates quiet ease; unlike that grill-centric grass yard, it is refined and subdued.

Following Auerbach into his home, Neville watches the man go about a well-rehearsed breakfast routine. Coffee and toast become, in Neville's retrospective narrative, a drama of socialisation. Neville is drawn to the way of life on offer but is also resistant, "off balance" (45): "A still life on the kitchen counter (…) two quarters of lemon on a ceramic tile decorated with a spiral, salt in a finger bowl. Ritual objects" (44). Auerbach's home is at once a still-life image and an active shrine. Neville knows enough of it to recognise its promise for someone like himself, but feels estranged enough to be fascinated. Importantly for an Anglophone novel with international sales potential, these emotions are mirrored in a certain subset of readers. I, for one, cast an uneasy glance at the Moroccan tile I use as a coaster at home while reading these lines, and my unease only intensified at the description of Auerbach's coffee ritual:

> I had never seen a cafetière before. He leaned on the plunger and gazed out of the window. It seemed to me that he was doing it in slow motion, building up tension in the room along with the mass of grounds in the bottom of the pot, drawing attention to the device. (44–45)

On my coaster was a mug; in the mug was some French-press coffee (oat milk, if you don't mind). The novel was interpellating me, its reader, disallowing certain detachments. If kitchen pots are the material world's point of entry into our bodies, the meticulous attention to coffee pots and salt cellars is Vladislavić's potentially unsettling point of entry into his readers' homes, the collapse of their impunity. Together with his coffee, Neville is being offered inclusion into a social group which may include the reader, especially if she is sitting with her book in a place where colonial settlers came to stay. And so, Neville's inculcation into white Anglophone South Africanness is, potentially, the reader's implication in one or more colonial histories of her own (Rothberg 2019). The tension "building up (…) in the room along with the mass of grounds in the bottom of the pot" emphasises the complex settler-colonial, late-capitalist structure wherein the exclusion of some—of most—from the "ritual" of the cafetière is synonymous with the growing gap between those who partake in the consumption (of coffee, for instance) and those who shoulder the burden of its production and who in their bodies pay its price (Vladislavić 44–45).

Older-Neville's ironic recollection of Auerbach's home is joined by a second irony, between the home's comforts, derived from its placement within a global social idiom of class, and its material reality, derived from place-specific structures of labour. African people are, of course, an integral part of the global network of consumption that the term "global middle class" connotes. But Vladislavić's text gestures towards an attitude prevalent in certain South African circles, which perceive being "worldly" and being "African" as antithetical. In Achille Mbembe's words, "that South Africa is not in a position to extricate herself from this continent is a source of great frustration" for many people, not all of them white; these people, he continues, do not see it as self-evident that "our citizenship in the world is inseparable from our citizenship at home, in Africa" (Mbembe 2019). I suggest that the perception of a tension between Africanness and worldliness is not limited to political and economic isolationist politics but experienced as a lived anxiety tightly bound up with settler-colonial complicity. Whether in the hasty opening of South Africa's economy during the transition to democracy its non-white urban middle class inherited some of this sense of tension is a possibility to be addressed beyond the scope of this chapter.

After his return to South Africa following the democratic elections, Neville settles into the social placement offered by Auerbach. He never fully commits to it, with the characteristic threshold-lover's distrust for any such commitment, but his material choices tell the reader something his words qualify. A young reporter interviews him in his home following a small photography exhibition. In the kitchen, she asks about his artistic influences:

> 'Saul Auerbach,' she said, 'he was the reason you became a photographer.'
> 'No, we can't blame him for that.'
> 'But he influenced you.'
> I let the statement settle while I drove the plunger down to the bottom of the cafetière. (182)

The out-of-place teenager has become a full-grown Anglo South African—in action, if not in mind. Before Neville can wax poetic about Auerbach's photography, all the while making snarky mental comments regarding said waxing, his material environment confirms the young journalist's question. The workings of capitalism had changed in the years between the events; a concern for the invisible labour behind a cup of coffee has sported further class classifications, the offer of the beverage now followed by "It's Ethiopian, I believe" (168). The conversation pauses while "[Neville] ground the beans and [the reporter] read about the Ethiopian coffee-drinking ceremony on the package" (179). Details of production processes and of indigenous practices become parts of a particular kind of food, and like that food are utilised in interpersonal social signalling.

It is no coincidence that this Anglo South African atmosphere is marked by nouns neither English nor Afrikaans. Like Auerbach's Mediterranean abode, the *cafetière*, the *espresso*, and the *soufflé* point outwards from the local and present, underscoring the paradoxical social placement of colonial settlers in the late-capitalist moment. It is crucial to understand that apartheid was born out of the double frameworks of difference which defined Afrikanerdom: not black, not British. It was this latter negation which was arguably the most forceful in the creation of a sense of victimhood which allowed apartheid to consider itself, improbably, anti-colonial. Thus, while claims of Europeanness justified the minority's control over indigenous peoples through the paternalist logic of colonialism, claims of indigeneity and locality were and are crucial for a social group who intended to stay. White South Africans are, after all, African—they have no other homeland. This is not a theoretical paradox but one which underlies lived experience. Apartheid South Africa's complex identity-naming system reflected this tension: "African" was never a term used to relate to the many black populations of the land, since the Afrikaan term Afrikaner was the very name the white Dutch descendants called themselves (Verwey and Quayle 2012, 555). Indeed, it was this assignation of locality to the Boers themselves which necessitated the invention of the derogatory name Bantu as a term for brown-skinned South Africans (ibid.).

Marlene van Niekerk's analysis of a culinary history of the Afrikaner reveals the same tension that I believe Vladislavić captures so well. Any attempt to describe a distinct and "pure" Afrikaner cuisine discernible immediately with the Dutch settlers' migration to the Cape betrays "an ideological agenda to ascribe an authentic European-rooted kitchen to the Afrikaner" (van Niekerk 2011, 67). Cookbook authors and scholars alike are then doomed to self-contradiction, describing this fare as authentic and ancient on the one hand and brought over from Europe on the other (van Niekerk 2011, 74). What I suggest is that while this tension is evident in Afrikaner history, it is nevertheless constitutive not of an ethnic identity but of a political positioning—namely, the settler-colonist's position within the political structure.

Therefore, when apartheid "sat down" as Vladislavić ingeniously describes it, Afrikaner and Anglo South Africans found themselves renegotiating both Europeanness and locality. Many chose to leave, seemingly electing one antipode of this tension and taking advantage of the privileges of British passports; yet even those who leave remain South Africans, if one is to believe a financial emigration firm which offers its clients, together with wealth management consultations, recipes for *Boerekos* (Afrikaner farmer's food): "In our haste to become global citizens", their website laments, "we've overlooked the foods we grew up with" ("Good Old Fashioned Boerekos? Yes, Please!" 2019). Others stayed, using an advantaged material position to make the most of the liberalisation of the domestic economy and to reposition themselves in the neoliberal order occasioned by the transition (Blaser and van der Westhuizen 2012).

Conclusion

Vladislavić's *Double Negative* thus presents us with an eating culture invested in references to locality and indigeneity and, conversely, allusions to global middle-class aspirations. I pointed out at the outset that if such a global class exists, African people of all ethnicities—and their food cultures—are surely already part of it. Why, then, does the careful dance of the *cafetiere* and the *soufflé* seem so important in the novel? The answer is that the tension the novel creates between "local" and "global" food rituals is reflective of the ironic tension of Europeanness and locality which is central to settler colonialism. European settlers, both Dutch and British, came to the southern horn of Africa under the veil of European superiority. The overwhelming economic advantages white South Africans still hold as a group cannot but be couched in, or at least refer back to, the same claim. Settling, on the other hand—disenfranchising native populations and building a life on the territory for generations—demands a rhetoric of nativeness, a relationship with the land which forecloses the possibility of "going back". This second part of the settlers' claim is staked through food practices which make use of local materials and flavours and which re-enact the takeover of land and the production of debris: individual soufflés on the one hand and braai and prawns on the other.

Bibliography

Avieli, Nir. 2013. Grilled Nationalism: Power, Masculinity and Space in Israeli Barbeques. *Food, Culture & Society* 16 (2): 301–320. https://doi.org/10.2752/175174413X13589681351458.

Beaudry, Mary C., Lauren J. Cook, and Stephen A. Mrozowski. 1996. Artifacts and Active Voices: Material Culture as Social Discourse. In *Images of the Recent Past: Readings in Historical Archaeology*. Rowman Altamira.

Blaser, Thomas M., and Christi van der Westhuizen. 2012. Introduction: The Paradox of Post-Apartheid 'Afrikaner' Identity: Deployments of Ethnicity and Neo-Liberalism. *African Studies* 71 (3): 380–390. https://doi.org/10.1080/00020184.2012.740882.

Du Toit, André. 2014. Victims of 'British Justice'?: A Century of Wrong as Anti-Imperial Tract, Core Narrative of the Afrikaner 'Nation,' and Victim-Based Solidarity-Building Discourse. In *Ten Books That Shaped the British Empire*, ed. Antoinette Burton and Isabel Hofmeyr, 112–130. Duke University Press. https://doi.org/10.1215/9780822375920-006.

Dubow, Saul. 2014. *Apartheid, 1948–1994*. 1st ed. Oxford: Oxford University Press.

———. 1989. *Racial Segregation and the Origins of Apartheid in South Africa, 1919–36*. London: Palgrave Macmillan Limited.

Ehrick, Christine. 2006. Beneficient Cinema: State Formation, Elite Reproduction, and Silent Film in Uruguay, 1910s–1920s. *The Americas* 63 (2): 205–224. https://doi.org/10.1353/tam.2006.0152.

Feldman, Allen. 2002. Strange Fruit: The South African Truth Commission and the Demonic Economies of Violence. *Social Analysis: The International Journal of Social and Cultural Practice* 46 (3): 234–265.

Flanery, Patrick. 2013. Double Negative by Ivan Vladislavić—Review. *The Guardian*, November 8. http://www.theguardian.com/books/2013/nov/08/double-negative-ivan-vladislavic-review.

Good Old Fashioned Boerekos? Yes, Please! 2019. *FinGlobal* (blog). December 27. https://www.finglobal.com/2019/12/27/boerekos-recipes/.

Hever, Hannan. 2007. *ha-Sipur ṿeha-leʾom: ḳeriʾot biḳoratiyot be-ḳanon ha-siporet ha-ʿIvrit [The Narrative and the Nation: Critical Readings in the Canon of Hebrew Fiction]*. Tel Aviv: Resling.

Karon, Tony. 2018. The Bittersweet (and Spicy) History of South African Cuisine. *Explore Parts Unknown* (blog), January 24. https://explorepartsunknown.com/south-africa/the-bittersweet-and-spicy-history-of-south-african-cuisine/.

Krog, Antjie. 2000. *Country of My Skull*. New York: Three Rivers Press.

Mbembe, Achille. 2019. Blacks from Elsewhere and the Right to Abode. *ZAM Magazine*, October 3. https://www.zammagazine.com/chronicle/chronicle-41/923-essay-blacks-from-elsewhere-and-the-right-to-abode.

Nobbs, E. 1905. The Importation from Abroad of Food-Stuffs Producible in Cape Colony. *South African Journal of Science* 4 (1): 70–86.

Osseo-Asare, Fran. 2005. *Food Culture in Sub-Saharan Africa*, Food Culture around the World. Westport, CT: Greenwood Press.

Ross, Robert. 1983. The First Two Centuries of Colonial Agriculture in the Cape Colony: A Historiographical Review*. *Social Dynamics* 9 (1): 30–49. https://doi.org/10.1080/02533958308458332.

Rothberg, Michael. 2019. *The Implicated Subject: Beyond Victims and Perpetrators*, Cultural Memory in the Present. Stanford, CA: Stanford University Press.

Scott, Rebecca. 2010. Meat My Hero: 'I Have a Dream' of Living Language in the Work of Donna Haraway, Or, Ride 'Em Cowboy! *Poroi* 6 (2): 3–14. https://doi.org/10.13008/2151-2957.1064.

Slatta, Richard W. 1983. *Gauchos and the Vanishing Frontier*. Lincoln: University of Nebraska Press.

Swart, Sandra. 2010. *Riding High: Horses, Humans and History in South Africa*. NYU Press.

Symons, Michael. 2007. *One Continuous Picnic: A Gastronomic History of Australia*. 2nd ed. Carlton, VIC: Melbourne University Press.

Truth and Reconciliation Commission Amnesty Hearing. 1996. Johannesburg. https://www.justice.gov.za/trc/amntrans/joburg/jhb.htm.

van der Vlies, Andrew. 2017. *Present Imperfect: Contemporary South African Writing*. 1st ed. Oxford: Oxford University Press.

van Niekerk, Marlene. 2011. The Eating Afrikaner: Notes for a Concise Typology. In *Reshaping Remembrance: Critical Essays on Afrikaans Places of Memory*, ed. Grundlingh and Siegfried Huigen, 63–78.

Veracini, L. 2015. *The Settler Colonial Present*. Palgrave Macmillan UK. https://doi.org/10.1057/9781137372475.

Versfeld, Marthinus. 1985. *Pots and Poetry*. 1st ed. Cape Town: Tafelberg.

Verwey, Cornel, and Michael Quayle. 2012. Whiteness, Racism, and Afrikaner Identity in Post-Apartheid South Africa. *African Affairs* 111 (445): 551–575.

Vladislavić, Ivan. 2013. *Double Negative*. London: And Other Stories.

Willard, Barbara E. 2002. The American Story of Meat: Discursive Influences on Cultural Eating Practice: The American Story of Meat. *The Journal of Popular Culture* 36 (1): 105–118. https://doi.org/10.1111/1540-5931.00033.

Wolfe, Patrick. 2006. Settler Colonialism and the Elimination of the Native. *Journal of Genocide Research* 8 (4): 387–409. https://doi.org/10.1080/14623520601056240.

Zreik, Raef. 2016. When Does a Settler Become a Native? (With Apologies to Mamdani). *Constellations* 23 (3): 351–364. https://doi.org/10.1111/1467-8675.12240.

Zycherman, Ariela. 2010. 'No Thanks, I Don't Eat Meat': Vegetarian Adventures in Beef Centric Argentina. In *Adventures in Eating: Anthropological Experiences in Dining from Around the World*, ed. Helen R. Haines and Clare A. Sammells, 203–222. Boulder: University Press of Colorado.

11

Sustaining the Memory of Colonial Algeria Through Food

Amy L. Hubbell and Jorien van Beukering

Introduction

In "Cookbooks and Aspects of Nationalism", Henry Notaker explores how difference became a part of defining nations and how defining a national cuisine became central to a country's identity. Notaker is especially interested in what happens when regional and national recipes are shared in different contexts: "When traditions crossed frontiers and entered new areas, they were absorbed by other cultures in an exchange of ideas about how to select, combine, and prepare different ingredients" (Notaker 2017, 246). Colonial Algeria provides a rich example of what happens when recipes cross borders, thanks to its long history of invasions, traders, migrants, and changing identities. The country was and continues to be inhabited by the indigenous Amazigh people, Jewish populations began arriving in the first century CE and Sephardic Jews

A. L. Hubbell (✉) • J. van Beukering
School of Languages and Cultures, The University of Queensland, Brisbane, QLD, Australia
e-mail: a.hubbell@uq.edu.au; j.vanbeukering@uq.edu.au

R. Ranta et al. (eds.), *'Going Native?'*, Food and Identity in a Globalising World,
https://doi.org/10.1007/978-3-030-96268-5_11

from Spain migrated there in the fourteenth and fifteenth centuries, fleeing persecution. Muslim Arabs came to Algeria in the mid-seventh century and today represent the country's major ethnic and religious groups. Parts of Algeria were conquered by the Spaniards in the late fifteenth century, and part was administered by the Ottoman Turks and known as the Regency of Algiers from the sixteenth to nineteenth centuries. During that time, the Ottoman Regency traded with Britain, the Netherlands, and France (McDougall 2007, 33). In the Napoleonic Wars (1803–1815), the region became a major trading partner and a source of grain for France. After French debts accrued and three years of French trade blockades on Algiers, in 1830 the French invaded and colonised the land. This was the beginning of the settler colony which brought waves of immigration from Spain, Italy, Malta, and even Germany in addition to the large number of French who were permanently implanted in the country until independence in 1962.[1] For 130 years, this complex mix of indigenous inhabitants and settler groups influenced each other's culinary habits through local ingredients and dishes combined with imported traditions.

Because of the many centuries of immigration and changing control of the lands in Algeria, it was often difficult to identify the indigenous people. The traditional owners of the land, the Amazigh, also known as Berbers, were repressed and persecuted, with Tamazigh not being officially recognised as a national language in the constitution until 2016 (Boukherrouf 2019, 10). The Algerian conquest to establish its boundaries continued from 1830 to 1956. While the Amazigh remained living in terrain that was difficult for the Europeans to settle, the Europeans flourished in coastal areas. In *Histoire de l'Algérie coloniale* (History of Colonial Algeria), Benjamin Stora calls the country heterogeneous, but also summarises Algerian demographics as Muslim, European, and Jewish with the colonial system "knitting together a chain of indigenous populations" (Stora 1991, 6). The relationship between ethnicities was simplified: "Il y a les 'indigènes' et les colons, les autochtones et les envahisseurs, les primitifs et les 'civilisés'" (17) ("There were the 'indigenous' and the

[1] Through the 1870 Crémieux Decree, French citizenship was given to European immigrants and Algerian Jews while excluding Muslims.

colonisers, the natives and the invaders, the primitives and the 'civilised'").[2] For European settlers in Algeria, the identity of indigenous populations was an amalgamation of Muslims, Berbers, and Sephardic Jews, as Stora sets out above, creating a duality of "us" and "them" upon which colonialism thrived. Any process of self-indigenisation for the Pieds-Noirs would not be directly related to the Berbers, but rather to any pre-existing ethnic group.[3]

The relationships between French and Muslims in colonial Algeria were often created on a power dynamic with the Muslims working for the French, especially in the later colonial years. Despite any social distance, however, certain things were shared. For example, in her book on European women in colonial Algeria, Claudine Robert-Guiard explains:

> Cette distance entre les êtres n'empêchait pas des emprunts de civilisation. Ces mêmes Européennes qui dénigraient les Arabes adoptèrent une partie de leur cuisine, à commencer par le fameux couscous, dont elles apprirent auprès de femmes arabes à rouler le grain. (Robert-Guiard 2009)
>
> This distance between beings did not prevent borrowings across civilisations. The same European women who denigrated Arabs adopted part of their cuisine, starting with the famous couscous for which they learned to hand-roll the grain from the Arab women.

Couscous is the most famous example of cuisine taken from Algeria that now thrives in France. But colonial Algerian cuisine was not only couscous: it was as diverse as the origins of the people who lived there. As Christophe Certain writes in the preface of his cookbook *Cuisine pied-noir*:

> Si chacun avait sa propre culture, les nouveaux arrivants ne tardèrent pas en revanche à échanger des recettes avec leurs voisins arabes, "français de France", grecs, libanais, maltais, italiens ou espagnols, donnant ainsi

---

[2] All translations are ours unless otherwise indicated.

[3] The Pieds-Noirs have been quick to point out that the Romans had also settled many parts of Algeria, leaving behind ruins in places like Tipasa, Timgad, and Guelma (Hubbell 2015a, 213). The diversity of populations in Algeria, the multiple waves of immigration, and the connection to ancient Romans served to further legitimise French presence in Algeria, as they contradictorily claimed the land belonged to no one and also to European ancestors.

naissance à un patrimoine culinaire d'une extraordinaire richesse. (Certain 2016, 7)

Although they each brought their own culture, on the other hand, the new arrivals did not wait to share their recipes with their neighbours who were Arab, "French from France", Greek, Lebanese, Maltese, Italians and Spaniards. This gave birth to a culinary patrimony that was extraordinarily rich.

Cooking traditions in colonial Algeria depended heavily on the communities in which they evolved, their ethnic and religious affiliations, social class, as well as regional influences on ingredient availability, and which local and migrant groups lived there. While the populations had more than a century to mix and meld their recipes, at the end of colonial rule, those recipes became fixed in place as sacred reminders of home.

Historian Jean-Jacques Jordi states, "ce n'est pas 1830 qui crée le Pied-Noir, mais 1962" (Jordi 2001, 114) ("it wasn't 1830 that created the Pied-Noir, but 1962"). After seven years of war, Algeria became independent, and nearly one million French colonial inhabitants were exiled to France where they established new communities. The diverse people formed a community and became collectively known as the Pieds-Noirs (lit. "black feet").[4] The traumas of war and exile left a deep mark on the community, binding them together. As it formed post-independence, the Pieds-Noirs' collective identity called upon a nostalgic version of Algeria as a lost paradise, filled with scenes of sun, beach, and harmony between the Europeans and the Arab majority (Hubbell 2015b, 8, 27). The homeland, to which they could not physically return, was recreated in literature and art, but also through their culinary practices as the community gathered annually in their new regions in France (Hubbell 2013). In exile, belonging to and coming from Algeria became central to Pied-Noir identity, and their memories strove to authenticate this "indigenous" identity. As Anna Johnston and Alan Lawson have pointed out, these authenticating narratives are designed to "legitimise the settler" in the colonised space (Johnston and Lawson 2000, 2005, 364). In other words,

[4] "Pied-Noir" has different origin stories: some claim the name originates from comparisons between French black military boots and bare Algerian feet, while others cite the pressing of grapes with bare feet to make wine.

the Pieds-Noirs cultivated attachments to Algeria as their identity became linked to a country that was now absent. Authenticating that relationship became essential.

As they created new communities in France, the Pieds-Noirs used their cooking traditions to feed their memories. Some of the diverse recipes were adapted or abandoned when products like *harissa* (hot chilli paste) were poorly tolerated in France, but others like couscous eventually became integrated into French national cuisine despite its North African origins (Wagda 1997b).[5] Creating an authentic couscous remains an important identity marker for many Pied-Noir families today. In France, associations sprung up representing the various regions in Algeria from where the Pieds-Noirs originated. Food played, and plays, an important role at Pied-Noir gatherings which focus on reconnecting with past traditions. Jean-Jacques Jordi explains the annual Pied-Noir Catholic pilgrimage and community gathering in Nîmes-Courbessac, which takes place

> …dans une ambiance de kermesse gigantesque où flottent les odeurs de brochettes et de merguez, [mais] ce qui compte reste la banderole que l'on agite au-dessus de sa tête indiquant le village d'origine suivi d'un point d'interrogation. (Jordi 2009, 127)
>
> …amid a giant festival atmosphere with the smell of meat skewers and merguez, [but] what matters are the banners that [the Pieds-Noirs] wave above their heads indicating their hometowns followed by a question mark.

These banners gather people from those same regions, while food reunites the people with their past. Food is central to the ambiance at these established gatherings as well as those planned post-COVID. For example, in the planning documents for the *Salon de la Culture et de l'Identité Pieds Noirs de Méditerranée* (Mediterranean Pieds-Noirs Cultural and Identity Expo), catered merguez and paella are at the top of the list.[6] Authenticity

[5] Wagda explains that merguez, which is typically Pied-Noir but was rarely served with couscous, has become part of the French national (and homogenised) dish of couscous (Wagda 1997b, 142). In 2016 couscous ranked among France's top ten favourite meals despite its Berber origin. Couscous' appropriation by Pied-Noirs and the French illustrates a particular forgetting of Algeria's colonial and pre-colonial past. This forgetting can be linked to colonial nostalgia (Lorcin 2013).

[6] Document sent to Amy L. Hubbell on 21 January 2020 from Mr Philippe Ruiz. "TR: 6 Manifestation 24-25 et 26 Juillet 2020 Barcarès". The event has been postponed to 2022 due to the

in both memories and cooking has become highly valued as the Pieds-Noirs attempt to preserve their memory of Algeria for their children, grandchildren, and themselves. In the settler-colonial context, culinary authenticity has come to mean a connection to Algeria and incorporates influences from all the different population groups that existed within Algeria: Amazigh as well as Arabic, Jewish, and Mediterranean.

Since the early twentieth century when Marcel Proust wrote about his experience of dunking a *madeleine* in his tea and being transported back to childhood, the connection between food and bodily memory has been well developed in French literature (Proust 1987, Holtzman 2006, 362). The recreation of the tastes and odours of past culinary experiences revive a connection to feelings entwined in those foods. A famous example of this in Pied-Noir literature comes from Marie Cardinal's *Les Mots pour le dire* (The Words to Say It 1975). When remembering a foundational moment in her childhood in which the young Maria feels ecstatically connected to her Algerian servants, she describes *méchoui* at a harvest festival:

> Une ambiance frémissante, comme des ailes de libellule. On préparait en sourdine la fête des vendanges. Il allait y avoir d'abord le couscous et le méchoui. Déjà les fosses étaient creusées et le bois préparé pour y faire de la braise. Les moutons dépecés, embrochés, empalés sur des pieux dressés contre le mur de l'entrée, attendaient de rôtir… Les femmes caquetaient autour du couscous qu'elles préparaient dans la cour. Elles étaient excitées. …Je me sentais légère, je restais avec les femmes à grignoter des raisins secs et des amandes grillées. (Cardinal 1975, 110)

> The atmosphere quivered like dragonfly wings. Preparations for the feast of the grape harvest were underway. First would come the couscous and the barbecued mutton. The pits were already dug and the wood gathered to make a brazier there. The mutton cut up, skewered and impaled on stakes, leaned against the entrance wall, ready to be roasted, any number of them! The women gossiped over the couscous as they were preparing it in the courtyard. Excitement was in the air. …I felt gay among the women, nibbling on raisins and toasted almonds. (Cardinal 1983, 130)

COVID-19 pandemic.

The event is recounted across several of Cardinal's texts because it represents the moment she felt most connected to the Algerian heritage. The narrator in *The Words to Say It* has been, like the author, exiled from her homeland and uses memory (and in this case memory of food and cooking) to reconnect with her homeland after the trauma of exile.

In addition to savours invoking memory, the cooking process connects one to family traditions and memories of the kitchen "back home" often come to the fore when reviving these culinary experiences. Jon Holtzman's definition of memory in "Food and Memory" includes "events that subjects recall or emotionally re-experience, the unconscious (perhaps embodied) memories of subjects, how a sense of historicity shapes social processes and meanings, nostalgia for a real or imagined past, and invented traditions" (Holtzman 2006, 363).

In this chapter, we examine works that are meant to sustain a dying tradition of colonial cooking as well as recreate those real and imagined pasts and invented traditions. Some cook to preserve traditions, some seek to solidify the tenuous affiliation with their ancestors in Algeria, and others simply want to relive a bodily memory of wholeness. We explore the representation of memory through cookbooks and websites intended for Pieds-Noirs and their descendants and recipes that transmit culture and tradition in both static and variable ways. Recipes for Arabic and Amazigh Algerian foods such as couscous, *chorba*, *méchoui*, and *mouna* and European foods like *creponet* (also spelled *créponné*), *macaronade*, merguez, and *calentica* are central to these works. All these foods are now considered Pied-Noir cuisine; traditional Algerian foods have been coopted into the Pied-Noir identity to validate the group's "authentic" ownership of and closeness to Algeria. Consequently, "authentic", "real", or "original" recipes are essential as individuals seek to revive and legitimise their own experiences in Algeria.

Cookbooks and cooking websites often have highly aesthetic representations of food to evoke nostalgia for what once was. While a recipe may focus on exact and precise ways of creating a dish, the static and fixed version of food may ensure its own disappearance when ingredients become unavailable or no one is left to demonstrate a certain technique. However, the practice of making good food and following recipes is far from static, and the acts of eating, sharing a meal, and conversing at the table are dynamic opportunities for change. The elements and methods used to produce a food are never stable or exact and the variability that enters into them is one that evolves in a healthy way.

In the sources examined, we identify competing strategies for maintaining memory through food. These variably rely on concepts of authenticity, originality, and singular rights to memory, but an undeniable realisation that foods belong to a community and that there is diversity is simultaneously intrinsic to recipes, cooking, and also eating. We explore how nostalgia functions in these texts using Svetlana Boym's concepts of restorative and reflective nostalgia (Boym 2001). We read restorative nostalgia as an unhealthy effort to restore or recreate the lost home, causing the individual to increasingly live in the past, which is an imagined homeland that cannot be "brought back" or restored. Boym's reflective nostalgia, however, accepts fragmented and disjointed, critical memories within it which we view as more productive as they allow for change. This chapter considers the ways in which foods attached to Algeria are preserved and fixed in time and how change is introduced and accepted in these recipes. In both blogs and cookbooks, the community's desire to sustain the memory of a fading past becomes central.

Cookbooks

Cookbooks are often full of enticing images and family, community, and national histories which contextualise and describe the origins of ingredients and dishes alike. While some individuals may possess family traditions written onto recipe cards and passed down through generations, for those who fled Algeria after the war, the recipes might only have been stored in their memories. In this section we examine three of the most popular and readily available Pied-Noir French-language cookbooks, one of which was published in 2001 and two in 2012.[7] These are Christophe Certain's *Cuisine pied-noir* (Édisud, 2001, reprinted in 2016), Brigitte

[7] An English-language cookbook, *Algerian Recipes: From Algiers to Constantine, Taste of All Algeria, in One Easy Algerian Cookbook* (2020) was consulted for reference. It has no individual author and contains no background stories. Apart from the beautiful colour cover, there are no images except for black and white clipart.

Erholt's *Mémoire et cuisine des pieds-noirs: la Nostalgérie* (Équinoxe 2012), and Léone Jaffin's *Algérie aimée: mes souvenirs et 222 recettes de là-bas!* (Le Courrier du livre 2012). Of the three, Jaffin's and Certain's cookbooks remain available for sale on Amazon.fr. Each of the books presents a different approach to the cuisine of colonial Algeria, both tightly bound to memory (Jaffin) and objectively distant from it (Certain). Erholt presents a more comprehensive (yet smaller) work that ties community stories into the recipes with beautiful images.

Originally published in 2001, Christophe Certain's *Cuisine pied-noir* presents his version of Pied-Noir cooking in a stark, straightforward text which presents a title, ingredients, instructions, some variations, and very occasional commentary. Though contained within a colourful cover, the volume is printed in black and white with clipart rather than photography, the recipes are categorised by course, and the book is dedicated to Certain's grandmother. In the foreword, Certain situates his authenticity within the story of his grandmother, born in 1914 in Bab El Oued (Algiers), Algeria, and underscores the diverse ethnic backgrounds and cooking traditions for those living in the colony (Certain 2016, 7). Certain describes his grandmother's home, stereotypically full of "les carpettes ornées de chameaux, les plateaux en cuivre, les photos de famille devant des palmiers…" (7) ("carpets decorated with camels, copper trays, family photos posed in front of palm trees…") but in contrast to his "banal" experience of being born and growing up in Nantes. Along with stories of Algeria told by his grandparents, these culinary traditions were passed down to him as post-memory (Hirsch 1997), and thus, his recipes do not contain detailed background, historical material, or communal and regional memories.

Certain's goal is to provide "authentic" and "flavourful" recipes (9) for classic Pied-Noir dishes such as couscous, for which he gives three different recipes including his grandmother Marcelle's, his aunt Georgette's, and a more unusual version with grouper. Certain points out that while couscous has become a French national dish, it is not often cooked properly or well. While his recipes remain mostly technical, Certain adds some personal advice about adding *harissa* (for which he also provides recipes) and where (in Paris) he had eaten the best couscous outside of his grandmother's home (54). The term "couscous" stems from the Berber *kuskus*; it most likely originated from the Sahara and/or the Sahel and is now widespread across the Maghreb, although the grain used across the region varied (millet or fonio in the Sahel; soft wheat or durum wheat in the Sahara and northern Africa) (Gast 2010, 70–73, 75). When the Pieds-Noirs arrived in France in the 1960s, they brought couscous with them, and it is now an industrialised product and one of the 'must have' foods intrinsic to Pied-Noir identity (Gast 2010, 77; Wagda 1997a, 166). Certain includes other classic recipes with Arabic origins such as *méchoui* (spit-roasted lamb) which he explains as "le repas de fête par excellence des Pieds-Noirs" (Certain 2016, 58) ("the ultimate celebratory meal for the Pieds-Noirs"). Scarce historical or memorial information is provided in this book which is restrained to factual cooking advice and some personal cooking tips. Certain exceptionally provides background rather than recipes for vegetables such as artichokes and Swiss chard and the use of rice in Algeria (influenced by the Spaniards and Italians [65]).

In a vastly different approach, Léone Jaffin proposes a personal memorial journey through 222 recipes from *là-bas* (back there). In the foreword, famous Pied-Noir actor Robert Castel cites Jaffin's love and "nostalgérie" as central to her work (Jaffin 2012, 8). Nostalgérie is the neologism coined by poet Marcel Fabri in 1938. The term is particularly used among the Pieds-Noirs for the longing they felt for Algeria (Hubbell 2015b, 27). Jaffin identifies as an ethnic Berber Algerian Jew who had been living in the USA prior to her arrival in Paris in 1978 (Jaffin 2012, 11). Feeling dislocated in Paris, she decided to reconnect to her Algerian roots through cooking. Her introduction provides a historic overview of her community as well as black and white photos of Algerian Jewish families in the 1920s and 1930s. In exile, Jaffin explains, "C'est autour de la table, quand ils réunissent leur famille et leurs amis, qu'ils peuvent se croire encore en Algérie et qu'ils réussissent à faire perdurer leurs traditions" (Jaffin 2012, 22) ("It's around the table when they gathered with family and friends, where they could still feel like they were in Algeria, and where they succeeded at making their traditions live on"). That is what inspired her book. No restaurant could translate those childhood memories for her or for her community. She says even her European friends wanted to share in these culinary experiences, "Car cette cuisine était et reste toujours confidentielle" (23) ("Because this cuisine was and still remains a secret"). In this way she specifically sets apart Jewish Algerian cuisine from regular North African cooking, and she claims to have been praised by famous Algerian Jews such as Jacques Derrida and Roland Bacri (26). However, despite being a sub-group within the Pied-Noir community, Jewish Algerian cuisine has been subsumed into Pied-Noir identity, highlighting the complexity of both this identity and indigeneity within Algeria.

The book is organised by Jewish holidays (Hanukkah, Passover, Rosh Hashanah, etc.) and the foods that should accompany each celebration. Jaffin intertwines family memories, family photos, and the requisite recipes for making each feast a success. These include touching images including Jaffin and her mother cooking "galettes blanches" (245) ("white cakes") for Purim. The recipes themselves are numbered and grouped together after each feast and are credited to those who provided them (Aunt Olga, Aunt Germaine, my cousin Monica, Emma Bensaïd, Éva, etc.). While the names create a sense of individual ownership, they also make the text seem even more authentic, as passed down by loved ones and close friends. Among the many recipes are the obligatory multiple couscous recipes: Meat and Vegetable Couscous (211) and Sunday Couscous (213). Although most of Jaffin's recipes are particular to the Jewish traditions she wants to share, Pied-Noir classics such as *créponné*, *chorba*, *cocas* (similar to empanadas), and merguez are also included. Interestingly, the famous "salade juive" ("Jewish salad") is not one of the recipe titles included in Jaffin's book. Jaffin manages to create a book that is at once intensely personal and also representative of broad community traditions.

Brigitte Erholt's *Mémoire et cuisine des pieds-noirs: la nostalgérie* also presents a communal memory of food. This small but dense book is colourfully illustrated and provides histories of everything from colonial Algeria and individual cities to ingredients used in the recipes. Like many Pied-Noir books that share memories of Algeria, Erholt includes personal family stories which stand in for the collective. She does this by using images that resonate with the community: maps, orientalist paintings, black and white images, food tins, posters, and postcards. As Hubbell explains in "Viewing the Past Through a 'Nostalgeric' Lens", "these visual texts ... participate in a communal autobiography. The same sorts of images are stacked and layered upon each other, creating a protective wall around the community" (Hubbell 2011, 177). Erholt recounts early childhood memories specific to Algerian life like visiting the *hammam*, after which she and her mother would "déguster un créponné (sorte de sorbet au citron) ou un zlabia (patisserie orientale à base de semoule, de fleurs d'oranger et de miel) en guise de goûter" (Erholt 2012, 20) ("sample a *créponné* [a type of lemon sorbet] or a *zlabia* [an Arabic pastry made with semolina, orange blossom and honey] as a snack"). Other memories she intertwines with stories like those told in Albert Camus' *Le Premier homme* (*The First Man*) to authenticate them as part of the collective (23). Erholt also recalls the war and that her family removed her from Algiers to live with her grandmother in Blida where there were fewer bombings. Nonetheless, she recounts one particular scary but not deadly bombing which woke the neighbourhood, causing her grandmother to invite the neighbours to share a meal:

> En effet, ils avaient eu une peur bleue en pleine nuit et elle voulait leur donner un remontant et soigner quelques égratignures. Il n'y avait pas eu fort heureusement de blessés. C'était ce qu'on appelait alors 'les événements'. Les voisins en pyjama, ou en chemise de nuit, attablés dans la salle à manger racontaient comment ils avaient été réveillés en pleine nuit, que des cloisons s'étaient effondrées… (Erholt 2012, 24)
>
> Indeed, they were scared to death in the middle of the night, and she wanted to lift their spirits and take care of their scratches. Very fortunately, no one was wounded. That was what we called then "the events". Neighbours in pyjamas, or in nightgowns, sitting at the dining room table and telling how they had been awoken in the middle of the night, and how the walls had collapsed…

Like many Pieds-Noirs who were children during the war, Erholt explains that the explosions were so common that it was hard to realise how serious the situation was. Writing as an adult, she sifts through fragments of traumatic memory which would be common to other Pieds-Noirs of her generation. These juxtaposed traumatic and happy memories demonstrate what Marianne Hirsch and Léo Spitzer call "ambivalent nostalgia" or "layered memories of 'home' made up of nostalgic longing as well as negative and critical recollections" (Hirsch and Spitzer 2002, 257). Erholt does not dwell on the trauma, but it remains present alongside the beautiful images of colonial Algeria and the descriptions of food as a source of healing.

Erholt's approach to colonial cooking is predominantly narrative. For traditional courses like *La Kémia* (appetisers) or drinks like *l'Anisette* (similar to ouzo), she fills a page with descriptions and stories (Erholt 2012, 41, 43). The recipes themselves, however, are singular and to the point. She lists ingredients, gives instructions, and occasionally adds a note on origins or serving suggestions. Most recipes include nostalgic colonial images such as her recipe for "salade juive" which is accompanied by the painting *La noce juive* (Jewish Wedding) by Armand Assus (69). Erholt acknowledges some of the recipes are controversial as each family has its preferences, but she navigates this by giving the "most complete" version. As she rightly explains for the *Tchoutchouka* (with ingredients similar to ratatouille and also known as "la frita"), "Chaque famille a sa recette et bien entendu chacune détient la vraie!" (70) ("Each family has its own recipe and of course each has the real version"). Erholt hits the quintessential Pied-Noir recipes in her book: *macaronade*, *méchoui*, merguez, *mouna*, *soubressade*, paella, *créponné*, and five couscous recipes. Like Certain and Jaffin, she includes her mother's version as well as a "Couscous de Latifa" who is a family friend. Erholt writes that though the dish was different from her family's, "J'en ai gardé un tel souvenir qu'elle a bien voulu me donner sa recette" (133) ("I kept such a strong memory of it that she wanted to give me her recipe").

Food Websites

Arjun Appadurai wrote that cookbooks "appear to belong to the literature of exile, of nostalgia and loss" (Appadurai 1988, 8). Food websites are the natural extension of the traditional cookbook online and can also contain nostalgic elements that easily connect exile, migration, and memory. For migrants and their descendants in diaspora, the Internet allows exiles to interact with their diasporic community and reconnect with Algeria through the sharing of recipes and memories (Kissau and

Hunger 2010, 246–248). In doing so, the websites remind migrants of their irrevocable separation from the homeland and therefore hold conflicting memories of French Algeria, containing both healing (reflective) and unhealthy (restorative) nostalgia.

In this section we analyse a personal blog, *Oran3644 Pour que le souvenir reste* (Oran3644 So Memory Remains), and two community websites, *AlgérieMesRacines* (My Algerian Roots) and *DenisDar: Le site de tous les pieds noirs* (DenisDar: The Site for All Pieds-Noirs). All three sites appear in the top twenty results of a Google search for *cuisine pied-noir recette* (Pied-Noir cooking recipe). Of the three, only Oran3644 tracks site visits (at 30 June 2021, it had been visited 201,621 times) (Martinez 2006–2013, accessed 30 June 2021). AlgérieMesRacines and DenisDar are currently active, while the last post to Oran3644 is dated April 2013.[8] The websites represent recipes as *lieux de mémoire* (sites of memory) and store precious, nostalgic memories of colonial Algeria (Nora 1989).

Restorative Nostalgia: Culinary Recolonisation

AlgérieMesRacines is maintained by two cousins: Agnès, born in Algeria in 1965, and Yves, born near Oran. Both now live in France and have direct memories of Algeria (AlgérieMesRacines n.d., accessed 28 May 2020). AlgérieMesRacines gathers documentary evidence of life in French Algeria, like recipes and photographs, to preserve the memory of Algeria. It is a community of Pied-Noir families: members can register their family and add stories specific to their family unit.[9] These contributions are imbued implicitly and explicitly with nostalgia by virtue of the impossibility of return to the place and time evoked in the recipes, stories, and photographs. AlgérieMesRacines thus functions simultaneously as forum, community space, and online encyclopaedia of everyday life in (French) Algeria while also forming an online memorial to life as it was *là-bas*. At the time of writing, the site contains 168 recipes and 20,533 photos in

---

[8] Blog owner Valérie Martinez indicated in July 2020 that she may revive Oran3644. See comment on post "La Mouna".

[9] In January 2021, 6703 members and 1995 family groups were registered on AlgérieMesRacines.

the category "Algeria". Similarly, DenisDar collects photographs and digitised postcards. It also features Pied-Noir recipes and lists of Pied-Noir books, films, and songs with titles such as *Le pays qui n'existe plus* (the country that is no more) and *C'était un beau pays l'Algérie* (Algeria was a beautiful country) (DenisDar n.d., accessed 20 May 2020). This hoarding of Algerian memory is indicative of Boym's restorative nostalgia: a nostalgia which aims to recreate the home that has been left behind (Boym 2001, xviii). The posting of photographs, family stories, and recipes on the websites serves to commemorate and reinforce an idealised image of Algeria and the densely presented collections crowd out competing versions of the past.[10] This is particularly evident in each site's "Recettes" section.

DenisDar lists fifteen "bonnes recettes pieds noirs" (good Pied-Noir recipes) and describes these as belonging to the "tradition culinaire de l'Algérie pieds noirs [sic]" (culinary tradition of the Algerian Pieds-Noirs) (DenisDar n.d., accessed 20 May 2020). Among these are Pied-Noir favourites of both Arabic and European origins, such as *mouna*, couscous, *oreillettes*, and *montecaos*.[11] The recipes are simple, unaccompanied by photographs, descriptions, or personal memories as is common on culinary blogs. Rather, DenisDar simply lists ingredients and instructions. Hints about how its author remembers Algeria can nonetheless be gleaned. For example, the description accompanying the couscous recipe reads "La recette originale du couscous" (emphasis added; ("the original couscous recipe"), while that for *oreillettes* is "La bonne recette des Oreillettes" (emphasis added; "the correct recipe for *oreillettes*") (DenisDar n.d., accessed 20 May 2020). These descriptions assert that these recipes are the original and authentic methods of making these foods. Because the website lists only one recipe per dish and the recipes cannot be contested through comments (despite visitors being able to comment on

[10] On the hoarding of Pied-Noir memory, see Amy Hubbell, *Hoarding Memory: Covering the Wounds of the Algerian War* (2020).

[11] *Mouna*: sweet Easter bread topped with sugar, sometimes marked by a cross (reminiscent of British hot cross buns). *Oreillettes:* sweet pastries often flavoured with orange blossom, also called *merveilles, chiacchieres, khechkhach. Montecaos:* small, round cinnamon-sprinkled biscuits. These foods are claimed as typically Pied-Noir and both recipe and spelling are contested (*mouna/mona, montecaos/mantecaos*).

other parts of the site), the idea of a singular correct version emerges. For the anonymous author of DenisDar, the indisputability of the recipes on their website reconstructs Algeria as it exists in his/her memory. The disabling of comments implies that no variations to the recipes presented are permitted on the site, meaning that the author's memory of Algerian foodways is uncontested. In Boym's words, this static memory "protects the absolute truth" of the author's memory of Algeria (Boym 2001, xviii).

AlgérieMesRacines also contains static memory. In contrast to DenisDar, recipes are contributed by users rather than by the site owners and comments are enabled on recipes. This allows recipe authors' culinary memories to be contested; yet, the recipes are surprisingly static. The Courbin family's "salade juive" recipe has twenty-one comments prescribing variations: several commenters support adding cumin while others recommend grilling the vegetables (AlgérieMesRacines n.d., accessed 19 April 2020). Still, the recipe has not been edited to accommodate these suggestions: it remains without cumin or barbecuing. This is common on AlgérieMesRacines; although some authors engage with comments, their recipes are not altered because of commenters' suggestions. Further, while some commenters are open to trying new versions of a recipe, most are less tolerant and instead offer suggestions to "improve" the author's recipe according to how the commenter's own family prepared the dish in Algeria: "'Chez nous' on mettait une grosse pincée de cumin…" ("In our family, we put a big pinch of cumin"); "Mon père faisait griller les poivrons et les tomates" ("My father grilled the capsicums and tomatoes"); "moi je la mange fraiche avec des GRAINS DE RAISIN NOIR un régale" ("I eat it cold with PURPLE GRAPES, a treat") (AlgérieMesRacines n.d., accessed 19 April 2020). This right to memory—the unwillingness to alter one's own recipes and the paradoxical eagerness to contest the recipes of others—can be attributed to the nostalgic memories that individual Pied-Noirs associate with specific dishes. For commenters, foodways are a direct, tangible link to their Algerian past and contain family memories of colonial Algeria. Given that restorative nostalgia centres around what the nostalgic believes to be true, altering a family recipe would alter the "authentic" memory of Algeria embodied within it (Boym 2001, 41). By cooking from recipes that accompanied them, unaltered, in their exile, first-generation

Pied-Noirs recreate an Algeria that exists only in their memories and which can only be relived *through* memory.

Despite its present connotation as a sentiment with no negative consequences, nostalgia was initially identified as a disease (Boym 2001, 3–5). The connection to ill health persists in Boym's theory of restorative nostalgia. Although "restorative" sounds healing, its focus on the lost past can be unhealthy: by aiming to restore or recreate the lost home, one commits more and more of oneself to a past that cannot be recovered, becoming alienated from reality. This is evident in Pied-Noir Gérard Rodriguez's posts in the "Histoires" (stories) section on AlgérieMesRacines, where users share written memories of Algeria.

"Histoires'" most prolific contributor, Rodriguez grew up in Bône (present-day Annaba) before migrating to France and Canada.[12] His posts idealise his childhood in Bône in which food was central. In a memory titled "La Macaronnade à Sidi Djemil" (*macaronade* at Sidi Djemil),[13] Rodriguez describes a pleasant day of hunting, cooking, and eating with Arabs and Pied-Noirs on a Sunday in 1956. The two groups gather around a *macaronade*, a beef stew with pasta (typically macaroni) served with *galette arabe* (flat, round bread) cooked on fires on location. Rodriguez describes the *galette arabe*, prepared by Arabic women, as the best he has ever tasted (Rodriguez n.d., accessed 11 January 2021). In his memory the coming together around the *macaronade* of coloniser and colonised is peaceful and joyous. Yet as David Sutton remarks, "memories of food are prone to this kind of telescoping in which the idealised aspects of the past are what are best recalled" (Sutton 2008, 169). Rodriguez claims that the day centres around the preparation, cooking, and eating of the *macaronade* and that everyone partakes of the meal together as a communal foodway. However, except for the *galettes*, the Algerian women bring their own food and eat separately. (Rodriguez notes that as a child he did not understand why.) Further, when recounting the meeting of the Arab and Pied-Noir groups at the start of the day, he writes: "tout ce

[12] Rodriguez has published thirty-one posts across three series in *Histoires*.

[13] While Rodriguez spells *macaronnade* with two "n"s, it is usually spelt with one "n". We retain Rodriguez's spelling in his title of the memory and use the more common spelling when not quoting Rodriguez. Variations in spelling are normal as many of these words were transcribed from Arabic and Spanish, or orally transmitted.

monde là, se serrait la main, s'embrassait comme si, oui comme si cela se faisait depuis des siècles" (Rodriguez n.d., accessed 11 January 2021) ("everyone there was shaking hands and kissing each other as though, yes, as though that's the way it had been done for centuries"), despite the fact that the Algerian War had already begun two years prior. As Patricia Lorcin explains in "Imperial Nostalgia, Colonial Nostalgia", when Rodriguez portrays Pied-Noirs and Algerians sharing beef stew and bread in friendship, the memory "freeze[s] the coloniser and the colonised into images of good-natured cooperation, erasing any tensions and conflicts" (Lorcin 2013, 105). The war is almost entirely omitted from the memory and thereby from Rodriguez's truth of the past. Thus, a memory which is ostensibly about food becomes dangerous through its selectivity: descriptions of foodways are highlighted, while intercultural relationships are idealised (Lorcin 2013, 103).[14]

Not only do his nostalgic memories inaccurately represent the past, Gérard Rodriguez feels more alive in that past than in the present. In an emotional memory relating his family's Christmas and New Year traditions in Algeria, he laments, "je ne suis pas heureux, comme là bas [sic]: en somme je n'ai rien, car l'Algérie me manque terriblement" (Rodriguez n.d., accessed 11 January 2021) ("I am not happy like I was back then: in the end, I have nothing, because I miss Algeria so terribly"). The aromas of "mouton, les merguez, le couscous et la daube […] on me les à volés [sic]" (Rodriguez n.d., accessed 11 January 2021) ("mutton, *merguez* sausages, couscous and stew […] they stole them from me"). Rodriguez confesses that he feels like "a stranger" in the present and that the silent shadows of the past appear more real than the people bustling around him in the twenty-first century (Rodriguez n.d., accessed 11 January 2021). He is more present in the ghost-world of his Algerian memories and retreats into them to sustain himself, buoyed by the smells and savours that he relives. As the scope of the past constantly narrows in, this kind of restorative nostalgia causes an unhealthy disidentification with the present and ultimately alienates Gérard Rodriguez from the world around him.

---

[14] The memory also exhibits other characteristics of colonial nostalgia, like romanticising the colonial landscape.

Reflective Nostalgia: Healing Memory

Pied-Noir food websites can also sustain healing memories of Algeria. Typically, healing memories are associated with reflective nostalgia, which is defined by Boym as "[thriving] in … the longing" for the lost time, and yet questioning the "absolute truth" of restorative nostalgia (Boym 2001, xviii). Among Pieds-Noirs, reflective nostalgia can be linked to the younger Pieds-Noirs who were children at departure and the second generation, for whom nostalgic memory of Algeria is about identity: growing up with what Marianne Hirsch calls "post-memories" of a lost home, children of migrants maintain links with the Algerian past by searching for roots in the nostalgia imparted through their parents' memories (Hirsch 1992, 8).[15] Often, it is in food that the second generation finds a sense of belonging and a space where memory transmission can take place.

Blog Oran3644 is maintained by second-generation Pied-Noir Valérie Martinez. While Martinez's parents lived in French Algeria, their daughter knows the former colony only through post-memory, through "des mots tel que calentica, mouna, roïco" (Martinez 2006) ("words like *calentica, mouna, roïco*")—her blog contains recipes for these and *oreillettes, frita*, and *montecaos*. Like Christophe Certain, Martinez situates the authenticity of her recipes within her family's Pied-Noir history.

In her article "Food in Binary", Kerstin McGaughey argues that food bloggers actively select which foods they share on their blogs and that this curation of food is central to the construction and perception of blogger identity (McGaughey 2010, 75). Oran3644's recipe section is titled "Ma cuisine Pied-Noir", indicating ownership of both the foods and the memories (of Martinez's childhood) and post-memories (of Algeria) that the foods embody. Thus, via Oran3644, Martinez performs the identity of a second-generation Pied-Noir searching for roots in memories of Algeria, and food is one avenue through which she reconnects with her heritage. The blog design evokes colonial Algeria in a way that will resonate with the Pied-Noir community: its earthy tones—red, brown, ochre, yellow, orange—evoke the Algerian landscape, while the header

[15] Studies addressing nostalgia and identity in children of migrants include Maghbouleh (2010), Bowering Delisle (2013), and Hirsch and Spitzer (2002).

incorporates symbols of pioneer rural life in a warm region: a sunlit wagon wheel, an overflowing fruit bowl, an ox-driven cart.[16] Martinez declares her Pied-Noir roots ("Je suis fille de pieds noirs": "I'm a daughter of Pieds-Noirs") but continues to search for a concrete sense of belonging by revisiting family history and cooking beloved dishes from her childhood (Martinez 2006).

McGaughey notes that "foodways play a large role in a person's identity in a larger group" (McGaughey 2010, 79). For Martinez, the best way to find a place within the Pied-Noir community is through traditional Pied-Noir foodways. This has been a long and arduous process for her. In her *mouna* recipe, posted in 2007, Martinez wrote that she had been attempting to bake *mouna* since 1990, but this was the first successful result: "je suis très fière car c'est pour moi, un pas de plus vers mes origines pieds-noirs!" (Martinez 2007) ("I'm really proud because, for me, this is one step further towards my Pied-Noir origins!").[17]

By baking *mouna*, *oreillettes* and *roïco* Martinez pays homage to her roots, cements her inherited Pied-Noir identity, evokes nostalgic memories of her childhood, and connects to her post-memory of Algeria. Practising Pied-Noir foodways allows her to connect with that past and find a place for herself within it, bridging the gap between her heritage and her present self. Each of the recipes on Oran3644 is what Boym would call "cherishe[d], shattered fragments of memory" that Martinez's mother brought to France in exile (Boym 2001, 49). For Martinez, these fragments (the recipes) are not static and she does not adhere to a single "true" version of each recipe. Rather, she edits recipes as she cooks. Her *mouna* is a combination with her mother's recipe, which substitutes peanut oil in place of butter (Martinez 2007). Like the other examples in this chapter, this underscores the gendered transmission of foodways, in which women and particularly mothers are frequently "the bearers of

[16] Interestingly, while Martinez's family is from seaside city Oran, the design excludes blues and whites.

[17] *Mouna* has moved between Pied-Noir subgroups: originating from Oran, it has Jewish influences, but is now embraced as a quintessential Pied-Noir food. *Mouna*'s specificity to Oran, its Jewish influences, and subsequent adoption by the wider Pied-Noir community illustrate the complexity of Pied-Noir (culinary) identity, which absorbs elements of other cultural traditions and is shared across ethnic communities.

[…] culture" and tradition (Protschky 2009, 375). Martinez is equally flexible with *calentica*. She writes that some recipes advise resting the dough overnight, while others do not, but explains that for her own *calentica* she ultimately did not rest the dough but will next time determine which technique works better for her (Martinez 2009). This indicates an acceptance that there are diverse ways to prepare Pied-Noir foods, in contrast to the need for authenticity among first-generation Pieds-Noirs.[18] Indeed, Martinez's *oreillettes* recipe is prefaced by this statement highlighting the variations between recipes: "Nous avons sûrement tous une recette plus ou moins différente, portant le même nom, ayant les mêmes origines, mais elle reste celle que nous avons appris avec nos parents et nos grands-parents" (Martinez 2008) ("We all surely have a recipe that is more or less different, bearing the same name, having the same origins, but it remains the one we learned with our parents and our grandparents").

For Martinez, to maintain links with Algeria and the Pied-Noir past it is not necessary to adhere to a "true" recipe; diversity and flexibility are welcomed to achieve the desired product. This ability to embrace diversity is linked to the changed status of recipes among the second generation: here, recipes are passed on as traditions, while for first-generation Pied-Noirs a recipe's attachment is to "authenticity". This shift is linked to a changed perception of Pieds-Noirs: while among the first generation there is a strong discourse of victimisation, the second generation is more productive with their memories. Second-generation Pied-Noirs like Valérie Martinez embrace what they can glean from Pied-Noir traditions while acknowledging that there are diverse ways of being Pied-Noir and diverse memories of Algeria. This recognition allows the traditions to evolve and survive while celebrating a cultural heritage in Algeria.

[18] Christophe Certain's website "Cuisine Pied-Noir" approaches diversity differently, presenting thirty-three couscous, ten *mouna*, and two *oreillettes* recipes. Cuisine Pied-Noir was active 2003–2019, before being reactivated in 2021.

Conclusion

Through the process of collecting recipes, some authors and websites sift through available resources to present a solitary and indisputable version that appears authentic. This singularisation can be a problematic approach to memory, but it produces a recipe with instructions that can be followed. In other cases, the many versions of the past are layered, with images, histories, recipes, and memories entwined into one food (Hubbell 2020). In these versions, multiplicity is inherent, but the accounts become dense, and the recipes become impossible to follow due to the variants. In both cases we see that memory does not necessarily produce good cooking, and yet food is meant to evoke a specific connection to the past.

The restorative nostalgia predominant among first-generation Pieds-Noirs, fixating on "the way things were" and "how the recipe should be made", creates an idealised image of Algeria that cannot be measured or proven as an accurate representation of the colony. These recipes focus on singular authentic recipes, insisting that there is a right way to prepare a dish. Central to the collective Pied-Noir settler-colonial identity that formed in exile, they authenticate the Pieds-Noirs as belonging to Algeria and can be read as (unconscious) self-indigenisation. Conversely, younger Pieds-Noirs with confused memories of Algeria and the second generation who received the post-memory of colonial Algeria experience the reflective nostalgia that is less concerned with getting the dishes (and through them, the memory of Algeria) correct. Rather, they use cooking to evoke Algeria and to experience some connection to what they do not know absolutely. Reflective nostalgia acknowledges and engages with the diversity of dishes, recipes, and ways of making the same product, sometimes allowing for substitutions, replacements, and inventions so that the recipe may take life. These younger generations can embrace the diversity of Pied-Noir memories of Algeria in healing ways that are not possible with restorative nostalgia that clings to "the way things were".

Bibliography

AlgérieMesRacines. n.d. Accessed February 2020. http://www.algeriemesracines.com/.

Appadurai, Arjun. 1988. How to Make a National Cuisine: Cookbooks in Contemporary India. *Comparative Studies in Society and History* 30 (1): 3–24.

Boukherrouf, Ramdane. 2019. Le Berbère dans les textes des constitutions algériennes. Analyse des pratiques discursives. Approches discurvie et jurilinguistique. *Comparative Linguistics* 40. https://doi.org/10.14746/cl.2019.40.1.

Bowering Delisle, Jennifer. 2013. 'Iraq in My Bones': Second-Generation Memory in the Age of Global Media. *Biography* 36 (2): 376–391.

Boym, Svetlana. 2001. *The Future of Nostalgia*. New York: Basic Books.

Cardinal, Marie. 1975. *Les Mots pour le dire*. Translated by Pat Goodheart. Paris: Grasset. Reprint, 1983.

———. 1983. *The Words to Say It. Translated by Pat Goodheart*. Cambridge, MA: Van Vactor and Goodheart.

Certain, Christophe. 2003–2021. Cuisine Pied Noir. https://www.cuisine-pied-noir.com/.

———. 2016. *Cuisine pied-noir*. Le Rove: Édisud. Original edition, 2001.

DenisDar. n.d. Le site de tous les pieds noirs. Accessed February 2020. http://www.denisdar.com/.

Erholt, Brigitte. 2012. *Mémoire et cuisine des pieds-noirs: la nostalgérie*. Bologne: Équinoxe.

Gast, Marceau. 2010. In *Une hypothèse sur l'origine historique et culturelle du couscous. In Coucous, boulgour et polenta. Transformer et consommer les céréales dans le monde*, ed. Hélène Franconie, Monique Chastanet, and François Sigaut, 67–81. Paris: Karthala.

Hirsch, Marianne. 1992. Family Pictures: Maus, Mourning and Post-Memory. *Discourse* 15 (2): 3–29.

———. 1997. *Family Frames: Photography, Narrative and Postmemory*. Cambridge, MA: Harvard University Press.

Hirsch, Marianne, and Leo Spitzer. 2002. 'We Would Not Have Come Without You': Generations of Nostalgia. *American Imago* 59 (3): 253–276.

Holtzman, Jon. 2006. Food and Memory. *Annual Review of Anthropology* 35: 361–378.

Hubbell, Amy L. 2011. Viewing the Past through a Nostalgeric Lens. In *Textual and Visual Selves*. Lincoln: University of Nebraska Press.

———. 2013. (In)Edible Algeria: Transmitting Pied-Noir Nostalgia through Food. *Portal: Journal of Multidisciplinary International Studies* 10 (2): 1–18.

———. 2015a. Accumulating Algeria: Recurrent Images in Pied-Noir Visual Works. In *Framing French Culture*, ed. Natalie Edwards, Benjamin McCann, and Peter Poiana, 209–227. Adelaide: University of Adelaide Press.

———. 2015b. *Remembering French Algeria: Pieds-Noirs, Identity and Exile*. Lincoln and London: University of Nebraska Press.

———. 2020. *Hoarding Memory: Covering the Wounds of the Algerian War*. Lincoln and London: University of Nebraska Press. Book.

Jaffin, Leone. 2012. *Algérie aimée. Mes souvenirs et 222 recettes de là-bas*. Clermont-Ferrand: Le Courrier du livre.

Johnston, Anna, and Alan Lawson. 2000. Settler Colonies. In *A Companion to Postcolonial Studies*, ed. Henry Schwarz and Sangeeta Ray, 360–376. Malden, MA: Blackwell Publishing.

———. 2005. Settler Colonies. In *A Companion to Postcolonial Studies*, ed. Henry Schwarz and Sangeeta Ray, 360–376. Malden, MA: Blackwell Publishing.

Jordi, Jean-Jacques. 2001. *1962: L'Arrivée des Pieds-Noirs*. Condé-sur-Noireau, France: Autrement.

———. 2009. *Les Pieds-Noirs*. Paris: Le Cavalier Bleu.

Kissau, Kathrin, and Uwe Hunger. 2010. The Internet as a Means of Studying Transnationalism and Diaspora. In *Diaspora and Transnationalism: Concepts, Theories and Methods*, ed. Rainer Bauböck and Thomas Faist. Amsterdam: Amsterdam University Press.

Lorcin, Patricia M.E. 2013. Imperial Nostalgia; Colonial Nostalgia: Differences of Theory, Similarities of Practice? *Historical Reflections* 39 (3): 97–111. https://doi.org/10.3167/hrrh.2013.390308.

Maghbouleh, Neda. 2010. 'Inherited Nostalgia' Among Second-Generation Iranian Americans: A Case Study at a Southern California University. *Journal of Intercultural Studies* 31 (2): 199–218.

Martinez, Valérie. 2006–2013. Oran3644: Pour que le souvenir reste. http://oran3644.unblog.fr/.

McDougall, James. 2007. Algeria. In *Encyclopedia of Western Colonialism since 1450*, ed. Thomas Benjamin, 33–37. Detroit, MI: Macmillan Reference USA.

McGaughey, Kerstin. 2010. Food in Binary: Identity and Interaction in Two German Food Blogs. *Cultural Analysis* 9: 69–98.

Nora, Pierre. 1989. Between Memory and History: Les Lieux de Mémoire. *Representations* 26: 7–24. https://doi.org/10.2307/2928520.

Notaker, Henry. 2017. *A History of Cookbooks: From Kitchen to Page over Seven Centuries*. Berkeley: University of California Press.

Protschky, Susie. 2009. The Flavour of History: Food, Family and Subjectivity in Two Indo-European Women's Memoirs. *History of the Family* 14: 369–385.

Proust, Marcel. 1987. *Du côté de chez Swann*. Paris: GF Flammarion.

Robert-Guiard, Claudine. 2009. *Des Européennes en situation coloniale: Algérie 1830–1939*. Aix-en-Provence: Presses universitaires de Provence.

Stora, Benjamin. 1991. *Histoire de l'Algérie coloniale (1830–1954)*. Paris: La Découverte.

Sutton, David. 2008. A Tale of Easter Ovens: Food and Collective Memory. *Social Research* 75 (1): 157–180.

Wagda, Marin. 1997a. L'histoire d'une migration culinaire. *Hommes et Migrations* 1207: 163–166. https://doi.org/10.3406/homig.1997.2982.

———. 1997b. Le couscous: nouveau plat national du pays de France. *Hommes et Migrations* 1205 (janvier–février): 142–143.

12

The Predicaments of Settler Gastrocolonialism

Lorenzo Veracini

This book neatly identifies three moments in the evolution of settler colonial food and foodways: the foundation of settler colonial food cultures, a moment of transition between settler colonial rejection and appropriation of Indigenous foodways (a moment of settler indigenisation), and the trajectory of settler colonial food after decolonisation. It turns out that food and food cultures matter enormously to the settler colonial 'situation', a circumstance characterised by the domination of an exogenous collective that is determined to build a permanent home in the country of Indigenous peoples. As a settler colony *reproduces in the place* of an Indigenous society, food is crucial to its evolution because it is a prerequisite for its propagation. This is so both in the mechanical sense, in relation to the calories available to the individuals belonging to this expanding society, and in the sense of a developing settler commensality and collective identity. But there is another side to this nourishment:

L. Veracini (✉)
Department of Humanities and Social Sciences, Swinburne University of Technology, Melbourne, VIC, Australia
e-mail: lveracini@swin.edu.au

R. Ranta et al. (eds.), *'Going Native?'*, Food and Identity in a Globalising World,
https://doi.org/10.1007/978-3-030-96268-5_12

Indigenous starvation. If 'gastrocolonialism' exists (see Chao 2021), so does settler gastrocolonialism.

Controlling the Foodways Economy

Settler colonial studies has emphasised the territorial dimension of settler colonialism as a specific mode of domination, a point the editors of this volume emphasise when they note that 'the relationship between the settler and the indigenous is always in relation to issues of land and power' (this volume, 4). But settler colonial studies neglected so far to focus on what is to be done specifically with the appropriated territory. This book reminds us that land is appropriated to produce food and that food is consumed in culturally specific ways, ways that are specific to a colonising collective that is defined by its appropriation of land for the purpose of its social reproduction. Land is where food comes from and also where the settler sovereignty becomes territorialised. In other words, settler food is a prerequisite of settler colonialism but also its outcome, because to make settler food, the foodways of the Indigenous populations must be disrupted, whereas an ability to inflict damage on Indigenous foodways enables settler domination in the first place. This process eventually became a self-fulfilling prophecy (i.e., a prophecy only 'filling' for the settler self): one social body must starve so that the reproduction of another can be ensured. Manifest Destiny looks like famine if you take a better look.

Let me state the obvious, even if it is an obvious that bears repeating in the face of settler mythologies about 'fragile' Indigenous ecologies, and 'evanescent races': the Indigenous collectives the settlers encounter are not weak; it is their foodways that are deliberately targeted so that they should become so. Likewise, blaming Indigenous communities for their 'innate' dysfunction (e.g., for an alleged genetic predisposition to diabetes, see Hay 2021) is morally and scientifically wrong. It is not that settler colonialism occurred because the Indigenous peoples were somehow maladapted, but the other way around: Indigenous bodies have at times become weak because the settlers were able to control their access to Indigenous foodways. It is no coincidence that Thanksgiving celebrates the sharing of nourishing Indigenous food, while the settler rations that reach Indigenous communities are very rarely nourishing. Likewise, a classic trope of settler colonial imaginaries is that the settlers make the 'desert bloom', but this representation implies its exact opposite as well: settlers deliberately create food deserts. Canada's founding father John A. Macdonald's intuited this necessary prerequisite for settler colonial domination and insisted that the western Indian reservations that were meant to contain Indigenous people must be unable to feed its inhabitants (see Harring 2019; see also Daschuk 2013). This is a historical example, but the Israeli authorities were very recently counting the calories per person that could enter and be consumed in the occupied Gaza Strip (see Gross and Feldman 2015). Elsewhere, the highly processed and literally diseasing food that reaches remote Indigenous communities does not alter the food deserts that settler colonialism made (see, e.g., Belcourt 2018).

The reciprocal holds true, and the settler communities growing into settler societies routinely define themselves by the enormous amounts of food they can access, especially meat, and it is also not a coincidence that conspicuous consumption of summarily cooked introduced animals summarise the food cultures of most settler societies. In this sense, lamb, for example, is indeed the food of settler colonial Australia and certainly the food of 'Australia day', a celebration that Indigenous peoples are fiercely contesting for very good reasons (see this volume, 174; the 'date' of Australia's national holiday should be changed, but the meat of reference may be reconsidered too). Likewise, while Americans are the 'people of plenty', as David Potter argued in a seminal 1954 intervention that contributed to the establishment of American studies, and abundance obviously includes food, all the settler food cultures are defined by overabundance. As the author of the chapter dedicated to South Africa included in this volume remarks, 'it is the remains [of a gargantuan meal] which make some forms of eating performative of local settler identity' (this volume, 207).

Either Indigenous food sources are depleted or Indigenous access to food is restricted (or both). Ways in which Indigenous peoples were/are separated from their subsistence, a process that looks very much like what in Marxian terms is defined as primitive accumulation and is ongoing, include the enforcement of settler jurisdictions, and their borders and treaties demanding spatial containment away from ancestral provisioning grounds (many of the treaties that sanctioned the acquisition of settler prerogatives over Indigenous lands were tellingly about limiting Indigenous access to food in specific locations in exchange for the provision of exogenous food), assimilationist policies demanding that Indigenous women be confined in the home, strictly patrolled residential requirements for Indigenous peoples (i.e., boarding schools, reservations, and prisons), enforcing bans on ceremonies, gatherings, and social activities that included consumption of culturally specific food, and environmental contamination of Indigenous land.

Such a systematic restriction of access to Indigenous food can be deliberate, but at times the selective food deserts of settler colonialism result from ecological transformations that the settlers have relatively little control over. Introduced species transform ecologies, while introduced pathogens can compromise the operation of Indigenous economies. As the Indigenous collectives starve, the settlers face a new world of abundance. It is significant that the Aboriginal peoples of what would become Taiwan were classified by imperial bureaucrats as either 'raw' ('wild') or 'cooked' ('civilised'; see this volume, 67). Cooked food is ready to be consumed and assimilated; according to the metaphorical structures mobilised by these expressions, settlement itself could be seen as a type of food preparation. The editors of this collection note that food 'becomes an emblematic vehicle for settlers to write themselves into and shape the territory and history' (this volume, 5), but the opposite is true too, and such a writing is necessarily premised on an ability to regulate the Indigenous food worlds. Emblematically, as the authors of the chapter of Israel remark, the *za'atar* and the *'akub*, aromatic plants typical of Palestinian cuisine, are identified by the settler state as 'endangered species' to prevent Palestinians from consuming them. Ultimately, as the settler collective controls the food economy, it literally eats its Indigenous counterpart, a point encapsulated by Brazilian modernist Oswald de Andrade's *Manifesto Anthropófago* (1928). Brazil had cannibalised itself into existence, a circumstance that rings true for most settler societies.

Various processes operate sequentially or concomitantly and reinforce each other: the settler overconsumption of Indigenous food sources depletes them; the introduction of imported crops, animals, and farming techniques further displaces Indigenous food sources; the movement of Indigenous populations is restricted, thereby distancing them from their food sources (this happens whether they are resettled, as it happened in Peru, when the colonial regime resettled and concentrated Indigenous population in order to count, tax, convert, and mobilise forced labour, and whether the Indigenous collectives are confined in reserves on their own land, but significant sections of their landholdings are declared off limits; see this volume, 35). Finally, settler foods are imposed on Indigenous populations, thereby further removing them from their traditional food sources and cultural practices. Take the example of sheep in Australia again, as described in the relevant chapter included in this volume: as Aboriginal croplands were destroyed by introducing livestock and the health and social cohesion of Aboriginal communities deteriorated, Aboriginal people were forced to 'join or become dependent on the colonial system to survive'. In turn, '[s]eparation from local knowledge, skills, traditions, lands and/or language profoundly disempowered (and continues to disempower) Aboriginal people' (this volume, 186; on Australian Aboriginal food systems and their destruction to the point of unrecognisability under colonisation, see Pascoe 2018). Alan Greer's analysis of the dynamics of Indigenous dispossession (2012) outlines a similar dynamic for settler colonial North America. The cow and the bison share a history but in the long run do not share much ground (see Barnard 2020). And yet, and importantly, and this should be emphasised, Indigenous resistance always actively and effectively disrupts these dynamics, and Indigenous people creatively embrace new foods and new drinks and develop new patterns of consumption that ensure their community's social reproduction. The second and third parts of this book would not exist without this resistance and survival.

Settler Colonial Foodways and Food Cultures

The 'settler colonial foodways and food culture', the specific topic of this edited collection, require dedicated modes of analysis, modes that are distinct from those deployed in the analysis of the foodways that are typical of other modes of colonial domination (on colonial food, see Laudan 2015). Claude Levi-Strauss seminally concluded that a society's way of cooking expresses its foundational structures, and this intuition makes sense when appraising diverse modes of colonial domination too (1990). A heuristic typology of colonial foodways could thus be briefly summarised: in colonial settings, where the coloniser collective parasitically devours the fruits of colonised labour and endeavours to distinguish its foodways from those of the collectives it subjugates in accordance with what Partha Chatterjee has called the 'colonial rule of difference' (1994), a reliance of locally produced food and on native techniques of preparation (a reliance enhanced by an economy focused on the production of colonial commodities for export) would coexist with a symbolic reliance on imported food, food that must remain unavailable to the colonised. In these settings, hybrid and creole cuisines characterised by reciprocal borrowings will emerge. Colonial Peru fits in with this pattern, where 'Spaniards grew to appreciate and enjoy certain New World foods, such as chocolate, pineapples, chillies, and sweet potatoes', but attempted to retain access to 'their own staples—wheat bread, beef, pork, and wine'. The latter 'were considered the mainstays of a diet appropriate for a European' (this volume, 33).

The outline of Mexican cuisine that emerges from this volume's chapter on recent developments in the officially sanctioned codification of 'authentic' traditional Indigenous foodways aptly explores the tensions separating colonial foodways from settler colonial ones. The food prepared by a *mayora*, 'originally the wife or concubine of the Spanish American Hacienda owner, and therefore principal cook in the Estate', is not the food prepared by 'traditional cooks', the *Cocineras tradicionales*, who 'are often described as female cooks with indigenous ancestry'. The latter speak Indigenous languages and cook 'primarily for and within a First Nation community' (this volume, 86). And yet, the latter's 'ingredients, recipes, techniques and preparations have increasingly found their way into *mestizo* repertoires', ensuring the emergence of 'an overlapping mosaic of culinary cultures, rather than a rigid distinction between settler and native foodways' (this volume, 100). Then again, colonial and settler colonial foodways routinely mix, as outlined in the chapter dedicated to Taiwan chronicling the shifts in rice consumption from Indigenous rice cultivars to rice cultivars imported by the Han settlers and then to those initially developed and propagated by the Japanese colonisers, but then embraced by both the post-war Nationalist regime and then the post-reform supporters of Taiwanese distinctiveness vis-à-vis mainland China (who have also embraced Aboriginal food; see this volume, 253).

In settler colonial settings, on the other hand, a rejection of Indigenous foodways would coexist with a determination to appropriate them (while attempting to disguise this appropriation). Hybrid cuisines are less likely in these contexts, especially where the settler economies often focus on subsistence and where the settlers are mainly subsistence farmers detached from markets. In these settings, appropriation often predominates over incorporation. At times, as the chapter dedicated to South Africa included in this volume outlines, a colonial emphasis on provisioning ships on their way to the East Indies (with no incorporation of Indigenous foodways) was successively superseded by a new and more settler colonial stance, after the Afrikaner settlers migrated inland, and 'picked up indigenous foodways, such as the curing of spiced meat into *biltong* or *droe wors*, and grilling fresh meat on an open fire, or *braai*' (this volume, 204). Then again, the social dynamics shaping settler food cultures inevitably interact in complex ways with issues of gender, race, and class, as illustrated, for example, in the chapters dedicated to Hawai'i and South Africa included in this volume: nineteenth-century Hawai'i cookbooks were designed to enforce gendered norms, while the specific racial implications of the Afrikaner *braai* were as pervasive as unspoken.

If creole cuisines are stunted under settler colonialism, Indigenous and settler foodways persist in their opposition, even if settler colonial domination systematically targets the Indigenous modes of social reproduction surrounding food. The moral economy of the settler project often holds that the settler community should be able to sustain itself in unmediated ways. In relation to Indigenous food, thus, the settler is ambivalent. This can be seen in the US 'Federal diet' addressed in one of the chapters included in this volume: some Indigenous foodways were readily incorporated while others were rejected. Those that were incorporated—maple syrup and corn, for example—were de-Indigenised in a variety of ways, either by assuming that the cultivars had originated elsewhere or by developing alternative harvesting practices (those that are rejected remain linked to Indigenous modes of consumption and are repressed). While the dehumanisation of Indigenous peoples included their real or imagined foodways (i.e., alleged cannibalism or 'bug' eating), the settlers more often than not partake.

What Terry Goldie understood as related settler colonial stances vis-à-vis Indigenous ways, 'penetration' and 'embrace' (1989), are confirmed in the context of the settler colonial foodway economy, and while the settlers have at times rejected Indigenous food, they have also routinely stolen it. They ate it, even though it was not theirs, and then they ate it *as if* it was the traditional food of *their* community. This cultural appropriation helped them craft claims about their indigenisation, claims useful when arrogating a right to dominate the land to the exclusion of colonisers who remained in the old country. A triangular system of relationships and the need to craft different claims thus counsels the conspicuous consumption of different foods in relation to different audiences. The settler foodways must include Indigenous food, so that the settler community can differentiate itself from its metropolitan counterpart but must also transform the land so that it can produce and consume imported foodways so that the settler community can distinguish itself from the Indigenous one. Thanksgiving should then be associated with the Tea Party because an embrace of Indigenous food parallels a rejection of exogenous consumption. Together these events epitomise a settler colonial dialectic and its tensions: embrace and refusal. Many Zionists explicitly rejected traditional *Ashkenazi* food culture; the US 'Federal' diet eschewed imported commodities (indeed, I suspect that attempts to codify a Federal diet in the late eighteenth century were designed to counter and challenge an 'antifederalist' one, which probably looked a bit too indigenised as far as the New England elites were concerned—two distinct settler projects challenged each other for ideological supremacy over an idealised settler table).

A remarkable variety of settler colonial food cultures emerge from this dialectic.

Decolonising Food and Food Cultures

The editors of this volume ask: what 'does a process of culinary decolonisation look like' (this volume, 149)? It is a question worth asking, and since specific food cultures characterise the settler colonial situation, transformed food cultures will be crucial to decolonisation too. We should actively craft decolonial foodways, a decolonial diet rather than a Federal one. Rather than deliberately instituted food deserts, decolonising settler colonialism will be about commensality—a *buen vivir*. I envisage a post-settler Indigenous-led potlach (even though there are equally decolonising but somewhat less enticing alternatives, as I argue below). There can be entertainment in the ways we entertain each other, after all, treaty, the recognition of Indigenous sovereignty and prerogatives, and 'treat' share a similar etymology. We will not deny each other, but something fundamental must change: the settler will be a guest, not a host.

Hospitality is crucial to all sovereign claims, as Derrida seminally argued (2000; as one of the authors contributing to this volume notes talking about the Afrikaner *braai*, 'racial supremacy, and cordial hospitality at any cost' went together, 210). The settler's transition from illegitimate host to guest, a transformation that would inevitably involve a reconstitution of sovereign arrangements, is crucial and cannot be negotiated. And yet negotiated it is, in a bewildering variety of ways, ways that are thoroughly explored in this volume. This variety is indicative of what is at stake and how important this passage would be. The settler wants to be a host, an indigenised host, this is a marker of his dominion, and has been globally willing to consider alternatives, especially lately. But from reading this volume we learn that he has been considering the alternatives without addressing the fundamental issue: Indigenous food sovereignty.

The chapters contained in this collection explore many of these negotiations; indeed, one suspects that it is precisely to avoid addressing the fundamental issue of Indigenous food sovereignty that so many alternatives have been explored. 'Local' food is still settler colonial food. Local food is not Indigenous food. 'Local' as an adjective evades the Indigenous-settler divide; it is an escape, and calling imported crops 'local' (i.e., sugar and pineapple in Hawai'i) is mere marketing. As the settlers consume 'local' food, they do not relinquish control, and an embrace of locally produced over imported food, as it happened in Hawai'i following wartime rationing, won't do either. Likewise, 'authentic' food, food that emerges from a specific *terroir*, is still settler food. Some Israelis now appropriate Palestinians foodways, and at times they even say so, even though they always appropriated it (not daring to admit so). 'Arab' food is a denial of Palestine, a denial of *Palestinian* food; 'Galilean food' is a denial of Palestinian foodways. The settler can deny it is Indigenous food or can even recognise it is, but it makes little difference unless the settler relinquishes control. 'Artisanally crafted' food is settler food. 'Organic' produce is still settler food, especially if displacing Indigenous populations is a prerequisite of organic agriculture (on 'organic washing', and on the way Israeli 'organic agriculture' literally displaces Indigenous Palestinian villages in the West Bank, see Kotef 2020: 215–260). 'Sustainably produced' and 'regional' food are still settler food. Embracing 'cosmopolitan' and diverse 'multicultural' food, one thinks of Australia's 'promiscuous multiculinarity' (see this volume, 188), for example, is still a form of food dispossession.

Community gardens, farmers' markets, and school gardening programmes, even when they include Indigenous food, will not decolonise. Regenerative agriculture will not do it. Even 'kangatarianism' is not food decolonisation, even if it is a more sustainable option than the introduced alternatives, and so an embrace of 'bush food' is not a recognition of Indigenous sovereignty (on the contrary, as the author of the chapter of this volume dedicated to Australia notes, Indigenous food can become a 'fetish object' that promises *without delivering* 'an easy salve for colonial guilt'; this volume, 189). Australia's growing 'bush food' industry is a textbook example of appropriation and exploitation of Indigenous food systems (and intellectual property), and Indigenous peoples own and control only about 1 per cent of an expanding industry that takes advantage of the 'culinary, pharmaceutical, naturopathic, horticultural, and tourism opportunities of Indigenous bush foods' (Bohunicky et al. 2021: 150–151). Ultimately, even the embrace of an originally Indigenous cultivar can be a settler colonial move, as the chapter included in this collection on Macadamia cultivation and commercialisation in Australia demonstrates. The cultivar was reimported to Northern New South

Wales from Hawai'i—its embrace in the 1970s coincided with a new wave of settlement marked by a shift away from dairy production, but dependency on pesticides and soil erosion and a lack of genetic diversity affect this industry. Not even deference to Indigenous culinary traditions and an associated 'gourmetisation' of Indigenous food can be a decolonising transformation if it is not the Indigenous person that is inviting the settler at the table. A 'traditional' cuisine is thus not necessarily decolonial; Mexico's tradition of 'unequal and exclusionary *mestizaje*', which 'grew out of the post-revolutionary Republic as a signifier of national identity', is not decolonial. Moreover, as the authors of the relevant chapter note, 'by replacing commonplace terminology like "*mayora*" and "*comidera*", with the umbrella term "*cocinera tradicional*", official institutions further contribute towards the erasure of ethnic diversity' (this volume, 99). Food justice, the reclamation of everybody's direct access and control of food systems, is commendable, but it is not decolonisation (see Daigle 2019).

Decolonising foodways should acknowledge Indigenous 'Country'. Country must be acknowledged in non-tokenistic ways, and all commercial production must be owned and managed by Indigenous communities and proceed in consultation with them. The Indigenous communities must benefit from all commodification of Indigenous food: harvesting, production, preparation, and consumption must benefit Indigenous communities if cultural appropriation is to be rejected on moral and ethical grounds and if cheap and ultimately counterproductive settler 'moves to innocence' are to be avoided (see Tuck and Yang 2012). 'Culinary colonialism' (Grey and Newman 2018), the outright appropriation by settlers of Indigenous foodways, can only be superseded by way of Indigenous food sovereignty. Perhaps 'Indigenous' should become a denomination of origin that is controlled and guaranteed by Indigenous authorities endowed with substantive powers that include the ability to supervise and allocate naming rights, to ensure the sustainability of Indigenous production, and to protect the intellectual and cultural property of Indigenous collectives. Commodification should not necessarily be seen with suspicion; it may even become conducive to decolonial ends and benefit Indigenous harvesters, fishers, land carers and knowledge holders, processors, chefs, and consumers. The food will be nourishing in more ways than one.

The settler can leave, bringing his traditions with him, and then surviving, like the *pied-noir* community in France, on a bittersweet fare of remembered tastes, aromas, and assorted 'nostalgérie' (this volume, 230). This community uses its cooking traditions to feed its memories and sustain itself, but the settlers may also stay after decolonisation. This is what a decolonised settler-Indigenous convivium may look like: the settler becomes a guest on Indigenous land, and the Indigene becomes a host. But there is a previous passage before all this can happen. First, the Indigenous collective must ensure its reproduction, including its social reproduction. It must be allowed to cultivate its futurity. Only then it can entertain. Decolonisation is when the Indigenous collective is able to self-determine its modes of social reproduction, a passage that must include a full recovery of its food sovereignty, a sovereignty that nourishes. Only after it is re-established, this sovereign can host; without it, it is only settler foodways. No longer surviving on bitter settler fare, the Indigenous hosts will then be able to genuinely entertain. Or they will not; it will be up to them to decide.

Bibliography

de Andrade, Oswald. 1928. *Manifesto Anthropófago*. https://writing.upenn.edu/library/Andrade_Cannibalistic_Manifesto.pdf.

Barnard, John Levi. 2020. The Bison and the Cow: Food, Empire, Extinction. *American Quarterly* 7 (2): 377–401.

Belcourt, Billy-Ray. 2018. Meditations on Reserve Life, Biosociality, and the Taste of Non-Sovereignty. *Settler Colonial Studies* 8 (1): 1–15.

Bohunicky, Michaela, Charles Z. Levkoe, and Nick Rose. 2021. Working for Justice in Food Systems on Stolen Land? Interrogating Food Movements Confronting Settler Colonialism. *Canadian Food Studies* 8 (2): 137–165.

Chao, Sophie. 2021. Gastrocolonialism: The Intersections of Race, Food, and Development in West Papua. *The International Journal of Human Rights*. https://www.tandfonline.com/doi/abs/10.1080/13642987.2021.1968378?journalCode=fjhr20.

Chatterjee, Partha. 1994. *The Nation and Its Fragments*. Princeton, NJ: Princeton University Press.

Daigle, M. 2019. Tracing the Terrain of Indigenous food Sovereignties. *The Journal of Peasant Studies* 46 (2): 297–315.

Daschuk, J. 2013. *Clearing the Plains: Disease, Politics of Starvation, and the Loss of Indigenous life*. Regina: University of Regina Press.

Derrida, Jacques. 2000. *Of Hospitality: Anne Dufourmantelle Invites Jacques Derrida to Respond*. Stanford, CA: Stanford University Press.

Goldie, Terry. 1989. *Fear and Temptation: The Image of the Indigene in Canadian, Australian, and New Zealand Literatures*. Montreal: McGill-Queen's University Press.

Greer, Allan. 2012. Commons and Enclosure in the Colonization of North America. *American Historical Review* 117 (2): 365–386.

Grey, S., and L. Newman. 2018. Beyond Culinary Colonialism: Indigenous Food Sovereignty, Liberal Multiculturalism, and the Control of Gastronomic Capital. *Agriculture and Human Values* 35: 717–730.

Gross, Aeyal, and Tamar Feldman. 2015. 'We Didn't Want to Hear the Word "Calories"': Rethinking Food Security, Food Power, and Food Sovereignty—Lessons from the Gaza Closure. *Berkeley Journal of International Law* 33 (2): 379–441.

Harring, Sidney L. 2019. 'Shooting a Black Duck': Genocidal Settler Violence against Indigenous Peoples and the Creation of Canada. In *Civilian-Driven Violence and the Genocide of Indigenous Peoples in Settler Societies*, ed. Mohamed Adhikari. London: Routledge.

Hay, Travis. 2021. *Inventing the Thrifty Gene: The Science of Settler Colonialism*. Winnipeg: University of Manitoba Press.

Kotef, Hagar. 2020. *The Colonizing Self: Or, Home and Homelessness in Israel/Palestine*. Durham, NC: Duke University Press.

Laudan, Rachel. 2015. *Cuisine and Empire: Cooking in World History*. Berkeley: University of California Press.

Levi-Strauss, Claude. 1990. *The Origin of Table Manners*. Chicago: University of Chicago Press.

Pascoe, Bruce. 2018. *Dark Emu: Aboriginal Australia and the Birth of Agriculture*. Melbourne: Scribe Publications.

Potter, David M. 1954. *People of Plenty: Economic Abundance and the American Character*. Chicago: University of Chicago Press.

Tuck, E., and K.W. Yang. 2012. Decolonization Is Not a Metaphor. *Decolonization: Indigeneity, Education & Society* 1 (1): 1–40.

Index[1]

A
Aboriginal people (of Australia)
agriculture, 120
cooking, 192
cultivation, 121, 256
disempowerment of, 186, 251
dispossession of, 86, 174
economies, 111, 114, 116, 121, 185, 187
environmental destruction, 185
food experience, 192
food production, 185
injustice towards, 174
knowledge, 175
plant knowledge, 110–111, 113, 119, 121, 179, 189
prejudice towards, 175
relations with Australia, 116, 119–120, 122–123, 174–176, 180, 185, 188–192
relations with settlers, 111–112, 114, 175, 179, 185–186, 190
See also First Nation/s
Aboriginal people (of Taiwan)/ Austronesian aboriginal people
dietary habits, 67
displacement, 80
food and flavours, 78
land use, 69
migration to Taiwan, 71
plant knowledge, 66, 66n1, 68, 69
relations with the Dutch, 67, 71
rice growing, 70

[1] Note: Page numbers followed by 'n' refer to notes.

R. Ranta et al. (eds.), *'Going Native?'*, Food and Identity in a Globalising World,
https://doi.org/10.1007/978-3-030-96268-5

Aboriginal people (of Taiwan)/ Austronesian aboriginal people (*cont.*)
- shifting cultivation, 70–72
- Siraya, 71, 72, 72n7
- societies, 67
- tastes, 13
- tribes, 68–70, 72n7

Adoni, Meir, 158, 159, 161n9
Africa, 10, 101, 117, 215, 218
African food, 203–218
Africanness, 215
Afrikaans, 216, 217
Afrikaner
- Afrikaner-Anglo relations, 205, 206
- Afrikaner-British relations, 206
- *braai*, 203–214, 209n2, 218, 253, 255
- history, 217
- identity, 206, 217
- migration, 217

Afrikaner national party, 205
Agriculture
- aboriginal agriculture, 185
- agricultural settlements, 67, 152
- agriculture and trade, 67
- agro-colonies, 67n2
- Andean agriculture, 34
- Australian agriculture, 188
- Australian pastoralism and agriculture, 111
- British agriculture, 204n1
- and cattle, 122
- and diet, 25, 61, 68
- Iberian agriculture, 33
- in Israel, 150
- indigenous agriculture, 35
- industrialised agriculture, 136, 142
- Native American agriculture, 59
- organic agriculture, 256
- Peruvian agriculture, 25
- slash-and-burn agriculture, 70

Agro-capitalism, 15
Agro-foodways, 31
Akub, 148, 157, 165, 250
Alfalfa, 36
Algeria
- colonial, 17, 221–242
- cultural heritage, 87, 156, 241
- demographics, 32, 222
- food of, 17, 221–242
- French, 222–226, 222n1, 223n3, 225n5, 229, 234, 239
- French conquest of, 222
- heritage, 227, 239–241
- history of, 221, 222, 231
- independence, 222
- landscape, 239

Algerian Jews, 222n1, 231
Algerian War, 238
Algiers, 222, 229, 232
Allende, Salvador, 41
Alpaca, 30, 31, 34, 36n6, 39
Alpacas and llamas, 30, 34
Amazigh (Berbers), 221–223, 226
America/n
- citizens, 52, 130
- culture, 12, 48, 50, 57, 58, 62
- dietary habits, 49
- farmers, 52
- food, 51, 52
- identity, 57, 130
- national and racial aspirations, 48
- newspapers, 49, 50, 128

products, 51, 60, 138
society, 26, 40, 50, 51, 56–59
American Board of Commissioners for Foreign Missions (ABCFM), 128, 130
American First Nations, 190
Americas, 2n1, 7–11, 26, 28, 30, 32, 32n4, 37, 47, 53n5, 59, 60, 65, 88n1, 114, 183
Amerindians (Indians), 101
Anchovy, 28
Andean
communities, 29
food, 23–41
Highlands, 29
regions, 27, 34, 38
valleys, 28, 36
zones, 23, 30
Annaba (Bône), 237
Apartheid, 203, 205–207, 209, 210, 216, 217
Apple pie, 24
Appropriation/appropriating
cosmopolitan, 16, 150, 154–158
of food, 3, 15, 17, 123, 147–167 (*see also* Culinary appropriation)
of indigenous tradition, 6
of labour, 12, 27
of land, 12, 26, 27, 248
Arab
Arabness, 160, 161
culture, 15, 148, 149, 149n3, 151, 154, 156, 160–162, 165, 166
food, 15, 16, 148–151, 148n1, 149n3, 150n4, 154, 156, 158, 160–163, 165, 166, 256
heritage, 148, 157, 160
people, 166
region, 148, 151, 154, 161
states, 153
world, 159, 161
Arabic, 149n2, 154, 158–160, 163, 226, 227, 230, 232, 235, 237, 237n13
Arab-Jews, 16, 153, 158–160
Arab-Jewish food, 160
Archaeology, 30, 35, 70
sites, 94, 96, 113
Arequipa, 34
Argentina, 26, 208
Artisanal production/artisanal product, 148, 155
Ashkenazi Jew, 160
Assimilation, 26, 56, 57, 82, 160, 165
Atahualpa, 25
Australia
British colonisation, 174
food, 16, 109, 110, 176, 177n5, 183, 190
government of, 118
land use, 186, 188
national anthem, 174
Northern Rivers, 109, 110, 114, 121–123
plants, 114
relations with aboriginal people, 114
settler colonial heritage, 16, 175
White Australia policy, 180
Australia Day, 16, 173, 174, 174n1, 176, 183, 190, 191, 249
Australian Macadamia Society, 118, 119

Australian Rules Football (AFL), 112
Authenticity, 9, 14, 15, 87, 103, 104, 148, 155, 160, 162, 164, 179, 225, 226, 228, 229, 239, 241
Avocado, 12, 38
Ayllus, 34, 35
Aymara, 27

B

Baking, 95, 240
Barbeque, 17, 176, 191, 208
  See also Braai
Bean, 28, 29, 34, 39, 86, 155, 213, 216
Beef, 33, 37, 93, 122, 158, 181, 183, 184n10, 208, 237, 238, 252
Ben Moshe, Gal, 164, 167
Bhabha, Homi, 4, 48
Bible, 150, 153
Blogs, 165, 228, 234, 234n8, 235, 239
Boer War, 206
Boerekos, 217
Bolivia, 37
Botany, 55, 113
  botanical specimens, 112
Botany Bay, 113
Boym, Svetlana, 228, 235–237, 239, 240
Braai
  masculinity, 17, 205–212
  scene, 208, 209, 211, 212
  torture, violence, 209, 209n2, 210
  whiteness, race, 17, 205
  See also Barbeque
Bread, 33, 37–39, 51, 95, 100, 131, 136, 160, 163, 178, 183, 205, 235n11, 237, 238, 252
Breakfast, 49–51, 53, 209n2, 214
British Empire, 131
Bundjalung Aboriginal people, 15, 111
Bush food, bush diet, 180, 192, 256
Bush tucker, bush tukka, 180, 182, 192

C

Camelids, 30, 34
  See also Alpacas and llamas
Camus, Albert, 232
Canned food, *see* Canned meat/ Canned pineapple
Canned meat, 136, 140
Canned pineapple, 132–135
Capitalism, 8, 25, 26, 216
Cardinal, Marie, 226, 227
Castillos, Pedro, 41
Castle & Cooke, 133, 135
Catholic Ascension Day pilgrimage, 17
Cattle, 36, 122, 143, 182, 208
Centro de las Artes Indigenas (CAI), 95, 96
Certain, Christophe, 223, 224, 228–230, 233, 239, 241n18
Ch'arki, 31
Chavez, Hugo, 41
Chavez, Julio Cesar, 97
Cheese, 49, 52, 94n4, 148, 161, 164
Chef/s
  Arab, 163, 167
  celebrity, 23, 155

cosmopolitan, 167
female cooks, 14, 85, 86, 88, 90, 252
Hawai'i, 139–142
Jewish-Israeli, 15, 155–158, 161–167, 162n10
MasterChef (Israel), 157
Palestinian, 148, 149, 149n3, 156, 157, 162–164, 166
Peru, 23
restaurant, 149, 156–158, 192
Top Chef, 143
Chicha, a fermented corn beer, 29, 36, 39
Chicken, 137, 148, 154, 163, 184
Chile, 41
Chilli, 14, 33, 86, 159, 225
China, 49, 65, 66n1, 67, 67n2, 69, 71, 72, 74, 118, 136, 253
Chontales, 93
Christian feast, 38
Christianity, 35
Christmas, 183, 238
Chuño, 31
Class
lower class, 36, 102
middle class, 92, 93, 131, 134, 210, 214, 215, 218
upper class, 41, 141, 213
working class, 139–141
Coca, 30, 231
Cocineras tradicionales, 86–92, 103, 104, 252
Coffee, 17, 53, 55, 139, 204n1, 214–216
Colombia, 118
Colonialism, 3–7, 9–12, 15, 24–27, 32, 65–82, 100, 104, 111, 113, 121, 123, 127–144, 150–154, 162–164, 174, 176, 185, 186, 191, 211, 216, 218, 223, 248–250, 253, 255, 257
Columbian exchange, 36
Comidera, 89, 89n3, 90, 99, 257
Commodification, 11, 18, 87, 155, 157, 161, 164, 167, 257
Conquistadores, 32
Conservation, 119, 182
Conservatory of Mexican Gastronomic Culture (CCGM), 86, 89–91, 93, 96, 97, 99, 103, 104
Cook, James, 113
Cookbooks, cookery books
Algerian, 228n7, 229
American, 134
Australian, 114, 178–180, 183
celebrity, 156
community, 129, 138, 228
Hawaiian, 15, 127–144, 253
Israeli, 147, 156
19th century, 114
Palestinian, 167
Pied-Noir, 17, 227, 228
Corn, Indian corn, 28, 29, 30n3, 39, 51, 58, 59, 59n9, 100, 205, 253
Cosmopolitan/ism, 2, 10, 16, 17, 150, 154–158, 167, 213, 256
Cotton, 28, 71
Couscous, 17, 165, 223, 225–227, 225n5, 229–231, 233, 235, 238, 241n18
Covid-19, 11, 27n2, 92, 226n6
Creole, 7, 66, 81, 87, 88n1, 93, 101, 102, 143, 252, 253

Crown Land, 116
Cuban Revolution, 41
Cuisine
  Afrikaner, 204, 217
  Algerian, 223
  American, 12, 47, 48, 62
  Anglo, 131
  Arab, 158
  Chinese, 78
  colonial, 223
  commercial, 89
  communitarian, 91
  creole, 252
  European, 154
  French, 141, 180
  Hawaiian, 139
  hybrid, 66, 253
  Indigenous, 100, 102
  Israeli, 156
  Italian, 158
  Jewish Algerian, 231
  Levantine, 164
  Mestizo, 88
  Mexican, 14, 85–104, 252
  mongrel, 177
  national, 3, 47–62, 80, 123, 183, 221, 225
  native, 14
  Palestinian, 149n3, 250
  Pied-Noir, 227
  Poly-cuisine, 177
  postcolonial, 88
  regional, 99, 141
  South African, 205
  Taiwanese, 78
  Totonaca, 95
  traditional, 86, 87, 91, 97, 257
  white, 205
Culinary appropriation, 15, 147–167
Culinary colonialism, 123, 257
Culinary heritage/food heritage, 2, 93, 96, 155, 157, 160
Cultural appropriation, 148, 156, 165, 166, 254, 257
Cultural heritage, 87, 156, 241
Cuy (guinea pig), 12, 37, 38
Cuzco, 34, 39

D

de Toledo, Francisco, 34, 35
Decolonisation, 6, 10, 14–18, 142, 148–150, 154, 162–167, 191, 247, 255–258
De-indigenisation, 119, 253
Diaspora, 152, 233
Dinner, 50, 51, 101, 102, 140, 210, 212
Dispossession, 13, 15, 16, 26, 66, 111, 116, 153, 174, 175, 251, 256
Dole, James D., 132, 133, 135
Domestication
  camelid, 30
  highland, 28–30
  macadamia, 113
Dutch
  colonies, 204
  colonisation, 205
  farmers, 204
  missionaries, 71
  New Amsterdam, 58, 60
  relations with aboriginal people (Taiwan), 67
  settlers, 58, 217, 218
Dutch East India Company, 67, 71, 204

E

Eating, 24, 37, 57, 77, 114, 131, 134, 181, 188–190, 207, 208, 210, 212, 218, 227, 228, 237, 249, 254
Eating habits, 47, 58, 203, 210
Ecology, 11, 80, 118, 121, 123, 248
Emu, 178, 180–182
Encomienda system, 33, 38
Enslaved African labour, 38
Enslaved labour, 38
Environment, 28, 58, 71, 78, 80, 97, 109–111, 118, 119, 122, 123, 166, 175, 176, 179, 185, 208, 216
Epidemics, 25, 32, 34, 35, 37
Erholt, Brigitte, 229, 231–233
Ethnic food, 154
Ethnicity, 8, 11, 40n8, 86, 89–91, 101, 102, 191, 218, 222
Ethnopreneurialism, 87
Europe, 10, 28, 36, 37, 55, 60, 65, 101, 122, 151, 155, 178, 204, 213, 217
Europeanness, 16, 17, 210, 211, 216–218
Exile, 13, 17, 151, 152, 224, 227, 230, 233, 236, 240, 242

F

Farming/farmers, 6, 28, 29, 36, 38, 52, 52n4, 67, 69–72, 74–76, 79, 80, 85, 118, 122, 133, 141, 151, 166, 204, 217, 251, 253, 256
Federal diet, 8, 12, 47–62, 253, 254
Fertilizer/fertiliser, 69, 73, 74, 80
Fine dining, 86, 141
First Nation/s, 85, 86, 88, 98–100, 175, 180, 189–192, 252
First People, 2n1, 16, 174, 175
Fishing, 2, 28
Food culture
  Aboriginal (Taiwan), 66, 70, 81
  African, 218
  Afrikaner, 206
  American, 8
  Arab, 15, 148, 149, 149n3, 150n4, 151, 154, 156, 160–162, 165, 166
  Arab-Jewish, 160
  Ashkenazi Jewish, 152
  Asian, 65
  Australian, 16, 176, 177n5, 190
  contemporary, 14, 104, 109, 110
  cosmopolitan, 155, 156
  diverse, 15
  European, 153
  Hawaiian, 15
  hybrid, 11–14
  Indigenous, 2, 3, 6, 9, 18, 57
  Israel, 148–153, 150n4, 155, 156, 159, 160, 162, 164–166
  local, 80, 89, 148, 151, 153, 156, 161, 162
  Mexican, 89
  Mizrahi, 154, 159
  national, 2, 155
  native, 16, 17, 27, 37, 38, 57–60, 109–111, 114, 119, 121–123, 175, 178–180, 188–192
  Native American, 57, 58
  Native Hawaiian, 15
  Nikkei (Japanese Peruvian), 41n9

Food culture (*cont.*)
  Palestinian, 15, 148, 149, 149n3, 150n4, 151, 154, 156, 161, 162, 165, 166
  regional, 89, 148, 151, 153, 161
  settler colonial, 1–3, 5, 6, 8–11, 16, 18, 66, 80, 100, 103, 104, 247, 249, 251–255
  South African, 203–218
  Taiwanese, 13, 66, 68, 81
Food production, 15, 27, 35, 49, 111, 114, 121, 152, 155, 185
Foodscape, 1, 12, 15, 187
Food scholar, 148, 148n1, 149, 156, 164, 177, 188
Food sovereignty, 2, 3, 18, 110, 123, 255, 257, 258
Food system
  Aboriginal Australian, 185, 190
  Andean, 23–41
  control of, 26, 28, 32, 257
  Indigenous, 256
  Inka, 27, 28, 31, 33
  modern, 26
  Spanish settler, 23–41
Foodways
  African, 48n2
  African American, 48n2
  Algerian, 236
  American, 8, 24, 48n2, 134, 138
  Andean, 23, 39
  Australian, 16
  Buddhist, 135
  Chinese, 136, 140
  colonial, 58, 101, 252
  contemporary, 5, 25
  de-colonisation of, 18, 142, 144, 255, 257
  Filipino, 140
  Hawaiian, 15, 127, 128, 132, 134, 138–144
  hybrid, 12, 35–40
  Iberian, 33
  Indigenous, 2, 3, 6, 9, 12, 27–31, 40, 60, 81, 88, 204, 247, 248, 252, 253, 257
  Japanese, 135, 138, 140
  local, 66, 139, 151
  national, 12, 24
  native, 30, 100, 252
  Native Hawaiian, 139, 141
  Okinawan, 140
  Palestinian, 151, 256
  Peruvian, 12
  Pied-Noir, 240
  plantation labour, 139–141
  pre-Hispanic, 103
  primordial, 32n4
  regional, 23, 151
  rice-based, 139
  settler colonial, 1–3, 5, 6, 8–11, 18, 82, 100, 127, 144, 247, 248, 251–254, 257, 258
  South African, 17
  traditional, 88, 140
  working class, 139, 141
Food writer, 23, 156–158, 162, 166, 167, 191
Foraging, 2, 28, 95, 155, 157, 164
Formosa, 13
France, 17, 18, 60, 141, 213, 222–225, 225n5, 230, 234, 237, 240, 258
Fujimori, Keiko, 41

G
Gastrocolonialism, 18, 247–258
Gastrodiplomacy/culinary diplomacy, 14, 104
Gastropolitics/food politics, 12, 15, 24, 40, 41, 144
Gaza Strip, 151, 249
Gender, 8, 72, 86, 88, 102, 104, 130, 138, 253
Gender roles, 130, 138
Genes/genetic, 118, 120, 189, 248, 257
    genetic diversity, 118, 120, 257
Globalisation, 14, 15, 155
Goat, 34, 37
Gourd, 28
Greek, 224
Guatemala, 118

H
Haciendas, 25, 26, 27n1, 35
Han Chinese
    culture, 67
    immigrants, 68, 77, 80, 81
    migration, 67, 71, 72
    people, 66–69, 71, 72, 80
    settlers, 13, 66, 253
Hawai'i/Hawaiian
    American annexation of, 132
    American business in, 133, 134
    American interest in, 128
    American military in, 131
    feast, 15
    food and wine festival, 139–142
    food of, 15, 127–144, 256
    Ladies Society of Central Union Church, Honolulu Hawai'i, 130–132
    plantations, 110, 133, 137, 140–143
    Regional Cuisine, 139, 141, 143
Hawaiian Gazette, 130
Hawaiian language, 128
Hawaiian Pineapple Company (HPC), 133
Hawaiian Pineapple Growers' Association (HPGA), 133
Hawaiian Pineapple Packers' Association (HPPA), 132–135
Hebrew, 149n2, 151, 156, 157, 160
Herding, 28, 34
Heritage, 1, 2, 12, 14–16, 40, 60, 86–88, 90, 91, 93–96, 99, 100, 103, 104, 129, 143, 148, 151, 153, 155–157, 160, 161n9, 167, 175, 176, 192, 227, 239–241
Heritage tourism, 14, 91, 104
Hirsch, Marianne, 229, 232, 233, 239, 239n15
Hiura, Arnold, 136, 140
Hixon, Walter, 48, 128
Homeland, 6, 17, 60, 151, 152, 176, 217, 224, 227, 228, 234
Honolulu, 128, 131
Honpa Hongwanji Hawaii Betsuin, Hawai'i, 135–139
Hummus, 153n6, 160, 165, 167n12, 192
Hunting, 2, 6, 72, 237
Hybridity, 6, 10, 12, 13, 40, 86, 102, 104

I

Immigration, 13, 41, 41n9, 66, 70, 80, 136, 137, 160, 222, 223n3
Imperialism, 4, 26, 80, 174
Inca, 28, 31
Indigeneity, 15, 86, 91, 103, 104, 162, 163, 192, 216, 218, 231
Indigenous
  bodies, 248
  communities, 18, 87, 93, 94, 97, 189, 190, 248, 249, 257
  culture, 2, 6–9, 57, 99, 190–192
  ecologies, 248
  economies, 250
  food, 2, 3, 6, 8, 9, 18, 57, 110, 123, 157, 189, 249–251, 253–257
  gastronomy, 91, 103, 123
  knowledge, 6, 8, 103, 119
  land, 5, 18, 123, 250, 258
  origin, 48, 56, 99
  people/population, 2–4, 7–9, 12, 18, 26, 34–36, 40, 48, 48n1, 56–60, 62, 66, 68, 68n4, 121–123, 174, 190, 192, 216, 222, 247–251, 254, 256
  recognition of, 15, 174, 255, 256
  relations with settlers, 2, 4–10, 18, 66, 248, 254, 255, 258
  resistance, 251
  rights, 2, 18, 119
  social and political structures, 7
  sovereignty, 2, 3, 18, 123, 255–257
  terminology, 99
  tradition/s, 6, 192, 257
  tribes, 68n4
  women, 91, 100, 250
Indigenous animals/native animals, 182
Indigenous flora/indigenous plant, 14, 109, 110, 189
Ireland, 10
Irrigation, 29, 35, 69, 73, 74, 80
Israel
  citizens, 156, 160
  culinary scene, 162
  immigration, 160
  national symbols, 8, 151
  occupation of Palestinian territories, 153, 154, 166
  salad, 159, 160
  self-indigenisation, 150–154, 162
  settler colonial identity, 15, 149
  society, 149
Israel and Palestine, 8, 15, 147–167
Israeli food, 15, 147–167
Italian food, 154
Italian immigrants, 8
IUCN, 110, 113

J

Jaffin, Léone, 229–231, 233
Japan
  annexation of Taiwan, 13, 66
  colonial power, 13
  colonialism, 73, 78
  food, 41n9, 75
  migrants, 41n9, 66
  relations with China, 74
  relations with Taiwan, 74
  rule over Taiwan, 68, 69
  Sino-Japanese War, 68
Japanese labourer, 136, 137
Japanese women, 137, 138
Jappon tea, 9

Jewish food, 152, 159
Jewish holidays, 231
Jewish-Israelis, 15, 148, 151, 155–158, 160–167, 162n10
Jewish immigrants, 152
Jewish settlements, 154
  See also West bank settlements
Jews, 8, 151–153, 159, 160
  See also Arab-Jews
Johannesburg, 203, 211
Jordi, Jean-Jacques, 224, 225

K

Kangaroo, 16, 173–192
Kangaroo meat, 176, 181, 182
Kangatarianism, 182, 256
Kassis, Reem, 156, 161, 166, 167
Kattan, Faid, 149n3, 157
Keren Kayemeth LeIsrael-Jewish National Fund (KKL-JNF), 150, 150n5
Knafeh, 148, 161, 165
Kuomintang, 13, 68, 81

L

Labour
  Chinese, 41n9, 133, 136
  domestic, 129
  Filipino, 136
  Hebrew, 152
  Japanese, 136, 137
  Korean, 136, 137
  Okinawan, 136
  Palestinian, 152
  Portuguese, 137
Lamb, 9, 16, 38, 164, 173–192, 206, 230, 249
Lamington cakes, 176, 183
Land
  clearing, 11, 116, 118, 121
  of Cockaigne, 58
  grants, 33
  owners, 25, 27n1, 102
  purchases, 133, 150n5
  seizure of, 207
  tenure, 6, 104
Landscape, 6, 14, 24, 28, 34, 69, 71, 72, 80, 111, 114, 120, 134, 142, 212, 238n14, 239
Laudan, Rachel, 131, 138, 141, 183, 252
Lebanese, 224
Levantine, 149n3, 161, 164
Lifestyle, 48, 58n8, 117, 129
Lima, 23, 28, 29, 33, 34, 38, 39, 41n9, 155
Llama, 28, 30, 30n3, 31, 34, 36, 36n6, 39
Lunch wagons, 140

M

MacAdam, John, 112
Macadamia nut
  conservation, 119
  cultivation, 256
  ecology, 121, 123
  habitat, 116, 117, 119
  Hawaiian industry, 118
  history, 14, 15, 109–123
  industry, 118, 119, 121, 122
  orchard, 115, 116, 118–120
  production, 15, 110, 115, 117, 118, 121
  varieties, 118
Macaronade, 227, 233, 237

Maize, 12, 14, 29–31, 34, 37, 38, 40, 58, 94, 213
  See also Corn, Indian corn
Malawi, 118
Maltese, 8, 224
Mamdani, Mahmood, 5, 177
Maple sugar, 52, 52n4
Maple syrup, 12, 52, 55, 56, 56n7, 61, 253
Maple tree, 52, 52n4, 53, 56
Marngrook, 112
Masculinity, 17, 205–212
Mayoras, 14, 85–104
Meat
  canned, 136, 140
  eating, 37, 57, 181, 182, 207, 208
  imports, 131
  national celebratory, 16, 175, 185, 191
  settler colonialism, 174, 176, 185
  trade, 184
Meat and Livestock Australia (MLA), 173, 183–186, 183n9, 190
Méchoui, 17, 226, 227, 230, 233
Mediterranean, 155, 214, 216, 226
Memory
  bodily, 226, 227
  childhood, 226, 230, 232, 237, 239, 240
  and food, 226–228, 231, 237–239, 242
  fragmented, 228
  postmemory, 229, 239, 240, 242
  static, 227, 236
  transmission, 239
  traumatic, 232
Merguez, 17, 225, 225n5, 227, 231, 233, 238
Mestizo/Mestizaje, 7, 14, 86–88, 88n1, 91, 100, 101, 252
Metate, 90, 98
Mexican gatronomy, 87, 94, 96, 99, 103–104
Mexican Revolution, 88, 101–102
Mexico
  application to UNESCO, 85–86, 89–90, 94–96, 103
  government of, 93
Mezcal, 14
Michelin stars, 23
Middle East, 10, 16, 101, 153, 159, 161
Migration, 5, 11, 39, 71, 72, 118, 177, 217, 233
Milk, 9, 12, 39, 50–52, 58, 94, 137–139, 214
Millet, 13, 66–68, 68n4, 80, 81, 205, 230
Mining, 33, 41n9, 122
Missionary/missionaries, 71, 128, 130, 131
Mizrahi food, 153, 154, 159–161
Mizrahi Jews, 153, 159, 160
Mole, 14, 94, 97, 98, 102
Morales, Gloria López, 87, 89, 90
Mouna, 227, 233, 235, 235n11, 240, 240n17, 241n18
Multiculinarity, 188
Multicultural, 2, 16, 40, 155, 157, 175, 176, 188, 192, 256
Muslims, 222, 222n1, 223

N

Nakba, 150n5
National cuisine, 3, 12, 47–62, 80, 123, 183, 221, 225

National identity, 5, 16, 24, 47, 86, 88, 173–192, 208, 257
Native American
  culture, 48, 57, 58, 62
  people, 48n1, 56–58, 62
  society/ies, 27, 56–58
Native Hawaiian
  culture, 15, 131
  food, 15, 131, 141, 143, 144
  knowledge, 131
  population, 128, 135, 141, 144
Nativeness, 8, 16, 17, 210, 218
Neoliberalism, 87, 217
Neville, Lister, 203, 205–216
New South Wales, 14, 15, 109–123, 181, 256–257
New World food, 33, 35, 36, 252
New Zealand, 2n1, 118
Nikkei foods, 41n9
North Africa, 8, 153, 159
North America, 8, 9, 26, 48n1, 54–56, 58–60, 251
Northern Ireland, 10
Nostalgérie, 18, 230, 258
Nostalgia
  reflective, 18, 228, 234, 239–242
  restorative, 18, 228, 234–239, 242
Notaker, Henry, 221

O

O'ahu, 128, 133
Occupation, 13, 33, 39, 151, 153, 154, 161, 165–167, 174
Okinawa, 135
Old World food, 32, 34, 36–37, 58–61
Olive oil, 100, 151, 160, 162
Olive tree, 150, 151, 153
Oran, 234, 240n16, 240n17
Orange
  Jaffa orange, 151
  orange blossom, 232, 235n11
Organic
  farming, 118
  food, 155
  produce, 148, 256
Otomis, 97, 98, 100
Ottolenghi, Yotam, 156, 163

P

Palestine, 8, 15, 147–167, 256
Palestinian
  citizens of Israel, 156
  culture, 15, 148, 149, 149n3, 150n4, 151, 154, 156, 157, 161, 162, 165, 166
  dispossession of, 153
  food, 16, 148–150, 148n1, 149n3, 150n4, 152, 154, 156–158, 161–167, 256
  identity, 149, 151, 163
  influence on Israel, 150n4, 162
  Israeli government policies towards, 151, 154, 157, 166
  place names, 154
  politics, 15, 149, 162, 163, 165
  produce, 148, 152, 162, 164
  territories (occupied), 153, 154, 156, 161, 166
Pascoe, Bruce, 9, 175, 182, 185, 189, 190, 251
Passion fruit, 12, 38
Pennsylvania Gazette, 50n3, 52, 56

Peru
agrarian reforms, 26, 29, n27
geography, 24–25, 28, 30, 34–35, 40
Peruvian
diet, 24, 25
dinning, 24
food, 23, 25, 26
history, 5
politics, 25–27, 40
Peruvian Ministry for agriculture, 29
Pesticide, 110, 119, 120, 257
Petits blans, 8
Photography, 139, 211, 216, 229
Pied-Noir
food, 17, 227, 230, 239, 241
history, 239
identity, 17, 224, 225, 227, 230, 231, 239, 240, 240n17, 242
memory of Algeria, 231, 237, 239, 241, 242
traditions, 225, 227, 231, 240n17, 241, 258
Pig, pork, 12, 33, 34, 37, 38
Pilgrimage, 17, 225
Pineapple
canned, 132–135
exports, 133
plantation, 134, 136, 141
recipes, 134
Pisco sour, 12, 24, 25
Pizarro, Francisco, 33
Plantation, 15, 25, 41n9, 92, 110, 133, 134, 136–143
Plantation towns, 136, 140
Plantation workers, labourers, 25, 41n9, 133, 134, 136–141, 143
Potatoes
processing of, 31
varieties, 29, 31
Potosí, 25, 33–35, 38, 39
Poultry, 25, 37, 204n1
Prickly pear, 8, 151
Production zones, 39
Proust, Marcel, 226

Q

Qing Dynasty, 67–69, 71, 72, 76, 77
Quechua, 27, 29
Queensland, 110, 113, 115–118, 120, 178, 179
Quinoa, 29, 34
Quito, 28

R

Race/racism, 11, 17, 41n9, 87, 114, 129, 139, 248, 253
Rainforests, 30, 110, 112, 116, 119–121
Ramekin, 17, 203, 212–217
Recipe/s, 12, 18, 31, 40, 52, 53, 56, 86, 88, 92, 96–100, 103, 129–131, 134, 135, 138–143, 178–180, 183, 217, 221, 224, 225, 227–231, 233–236, 235n11, 239–242, 241n18, 252
Restaurant/s
Arab, Mizrahi, 154, 159, 160
Brut, 162–163
Cocina Chontal, 93
culture, 139, 155, 157
Dianah, 156
dinner, 102, 140

Habasta, 154–156, 158n7
Han-Japanese, 75
Hawaiin, 139–142
and hotels, 141, 142
Israeli, 156, 158, 165
Machneyuda, 155
Mexican, 89, 89n2, 91–95, 102
Tlamanalli, 93
Zippy's, 140
Rice
African, 65
Asian, 65
Basmati, 79
breeding, 73, 77–80
consumption, 65–82, 252
cultivation, 13, 66, 68, 70, 71, 73, 76, 79, 80
culture, 69, 81, 154
export, 74, 75, 77
farming, 70–72, 80
Indica, 65, 68–77
Japanese, 69, 73, 79
Japonica, 13, 65, 69, 70, 73–81
Jasmine, 79
Javanica, 13, 65, 68, 70–72
paddy, 68, 70
Penglai, 73–75, 77, 78
prices, 74, 75, 77, 79
production, 65–82
purity, 79
red, 68
upland, 13, 68–71, 80
varieties, 13, 50, 65, 66, 68–71, 73, 75–80, 82
vinegar rice, 75
Rice wine, 71
Rous, Henry, 111

S
Salad/s, 134, 140, 159, 160, 213
Santiago, 28
Self-indigenisation/self-indigenising, 2, 4–10, 16, 17, 82, 101, 150–154, 162, 175, 185, 204, 206, 223, 242
Settler colonial
domination, 5, 6, 80, 101, 247–249, 252, 253
exploitation, 4, 26, 190
identity, 2–4, 7, 15, 149, 203, 242
states, 2–4, 7
Settler colonialism
American, 100, 128
British, 218
Dutch, 218
French, 8
Israeli, 150, 151, 249
Japanese, 136, 137
Spanish, 12, 26, 100
Settler-indigenous relations, 6, 7
Sheep, 34, 37, 181–187, 251
Simeon, Sheldon, 142–144
Slavery
enslaved labour, 25
importation of slaves, 25
Slow food, 155
South Africa
democracy, 211, 215
post-apartheid, 209, 212, 215
South America, 11, 28, 32, 117
South Korea, 65
Spain, 222
Spanish America, 25, 33, 36, 37
Spanish conquest/invasion of Americas, 28, 87, 100

Spanish Crown, 25, 34, 38
Spanish settlement, 27
Squash, 14, 28
Stora, Benjamin, 222, 223
Street food, 89, 157, 205
Sugar, 25, 36, 39, 52, 52n4, 53, 55, 71, 77, 94, 133, 136, 137, 141, 204n1, 235n11, 255
Sugarcane, 36
Sugarcane whiskey, 36
Sustainable, 86, 122, 142, 182, 256
Swaziland, 118
Sweet potatoes, 33, 66, 66n1, 252
Symbol/symbols, 8, 9, 17, 101, 122, 150, 151, 153, 174, 185, 191, 208, 240

T

Tacos, 101, 102
Taiwan
  annexation by Japan, 13, 66
  colonization, 66, 81
  government of, 68, 69, 71–74, 76, 81
  history, 66
  identity, 13
  immigration to, 13, 66, 80
  national symbol, 78
  relations with aboriginal people, 250
  under Japanese rule, 69, 73
Taiwan council of agriculture, 76, 78
Taiwan department for agriculture, 67
Taiwanese
  culture, 78
  food, 13, 68, 70, 75, 81, 253
  society, 75, 78
Taiwanese Koshihikari, 13, 78
Taiwan's agriculture sector, 75
Taro, 13, 66, 66n1, 80
Tea, 9, 12, 53–56, 53n5, 56n6, 61, 77, 176, 177n4, 226, 254
Teotitlan, 89n3, 93, 94
Terroir, 2, 6, 99, 148, 151, 154–156, 160–165, 167, 256
Thanksgiving, 183, 190, 248, 254
Tlamanalli, 93, 94
Totonacas, 96
Tourism, 14, 24, 90, 91, 94–96, 104, 129, 134, 135, 256
Trade, 5, 30n3, 39, 67, 71, 74, 115, 117, 122, 184, 204, 222
Traditional cook, 14, 61, 86–88, 89n3, 90, 91, 95–97, 103, 233, 252
Transculturation, 7, 86, 87
Tree
  maple, 52, 52n4, 53, 56
  nut, 115, 122
  olive, 150, 151, 153
Tuntian system, 67, 67n2
Turkey (country), 59, 183
Turkey (poultry), 178

U

UNESCO, 14, 85, 86, 89, 90, 94, 95, 103
United States (US), 8, 47, 48n1, 49, 52, 53, 55, 90, 128–130, 133–135, 137–139, 253, 254
Urban/urbanity, 16, 17, 27, 36n6, 38–40, 69, 88, 101, 102, 155, 159, 181, 212, 215

V

Veracini, Lorenzo, 3, 4, 7, 18, 151, 177, 185, 188–190, 206
Victimhood, 18, 151, 216
Vietnam, 118
Vladislavic, Ivan, 17, 203–218

W

Wallaby, 178–181
Websites
 AlgérieMesRacines, 234, 234n9, 236, 237
 DenisDar, 234–236
 Oran3644, 234, 234n8, 239, 240
West Bank, 151, 157, 163, 166, 166n11
West Bank settlements, 158n7
Wheat, 12, 29, 33, 34, 37, 38, 40, 51, 58, 59, 71, 100, 155, 163, 186, 230, 252
White/whiteness
 dominance, 209
 identity, 206, 211
 liberals, 210
 middle class, 210, 214
 privilege, 206
 South African identity, 203
Whitewashing, 175
Widjabul Wia-bal people, 112
Wilderness/wild
 wild food, 120
 'wild' natives, 121
Wine, 24, 25, 33, 52, 71, 100, 102, 154, 158n7, 162, 213, 224n4, 252
Wolfe, Patrick, 4, 7, 32, 185, 186, 211
Women
 Arabic, 237
 and cookbooks, 130, 131, 138
 cooks, 39
 European, 223
 indigenous, 90, 91, 250
 Japanese, 137, 138
 native, 130
Wool/wool industry, 183, 184

Y

Yossefi, Yair, 163, 164, 167

Z

za'atar, 157, 165, 250
Zheng, Chenggong, 67, 71
Zionism, 151, 152, 159
Zionist leaders, 152
Zionist movement, 152
Zionist settlers, 8, 151–153
Zippy's restaurants, 140

FSC
www.fsc.org
MIX
Papier | Fördert gute Waldnutzung
FSC® C083411

Kolibri 360 GmbH
Ferdinand-Jühlke-Straße 7
99095 Erfurt, Deutschland
produktsicherheit@kolibri360.de